Mechanical Engineering Technology for CSEC®

Michael Barlow
Errol Clarke
Philbert Crossfield
Jerry Simpson

Contents

Access your support website for additional content covering the theory aspects of the syllabus here: **www.oxfordsecondary.com/9780198395614**

Introduction

Mechanical Engineering Technology for CSEC® a CXC Study Guide has been developed by experienced teachers and examiners, working closely with the Caribbean Examinations Council (CXC®) and focuses on the development of competencies. It concentrates on the areas of the syllabus that are most challenging to learn and are considered essential to the development of skills required by the programme and entry into the world of work.

The content has been constructed around workshop investigations to support the process of relating theory to practice and practice to theory. These are designed to align with the school-based assessment and allow students to review progress effectively. This gives the student a positive role in managing their own learning. Additionally, there are opportunities for students to use reflective techniques to identify what went well, what might have been done more effectively and how similar activities might be approached in the future (skill transfer).

The study guide and associated activities support a range of pedagogy to make learning engaging, interactive and efficient, leading to a deeper understanding. The range of pedagogy includes:

1. Assessment of and for learning
2. Cooperative learning
3. Differentiation
4. Embedding language, literacy and numeracy
5. Experiential learning
6. Learning conversations
7. Relating theory and practice
8. Using e-learning and technology.

Sections covering the theory aspects of the syllabus can be found on the support website at www.oxfordsecondary.com/9780198395614

Remember, where applicable, candidates who successfully complete the CSEC examinations in the Technical Syllabuses will receive two awards: the CSEC Technical Proficiency Certificate and a CVQ* (Caribbean Vocational Qualification) Statement of Competence.

We are confident that this book will provide students with the skills to succeed in their course of study and beyond.

*CVQ is the registered trademark of the Caribbean Association of National Training Authorities (CANTA)

1. Heat treatment methods

Option B Section 1.5 & 1.6

1.1 Concepts of heat production and transfer

Thermodynamics

Thermodynamics is a branch of physics concerned with heat and temperature and their relation to energy and work done. A thermodynamic system is one that interacts and exchanges energy (or transfers energy) with the environment around it.

Zeroth's law of thermodynamics states that *if two bodies are each in thermal equilibrium with some third body, then they are also in equilibrium with each other.* Perhaps put a little more simply: if systems "One" and "Two" are each in equilibrium with a third system "Three", it means they each have the same energy content as system "Three". Therefore all the values found in "Three", match those in both "One" and "Two".

The **first law of thermodynamics** is a version of the **law of conservation of energy**, which states that the total energy of an isolated system is constant; energy can be transformed from one form to another, but cannot be created or destroyed. For example, if you have a kettle of water in a kitchen at room temperature and then heat is add to the system (the kettle) the energy and temperature will increase, but some of the heat and energy will be released into the environment e.g. heating up the kitchen.

Note:

1. Energy (heat) cannot be created or destroyed, only transferred.
2. Not all heat can be used by a system to do work, some is always "lost" to the surrounding environment.
3. When two or more bodies are in a closed system, heat from the hotter body will pass to the cooler body until they are both the same temperature, because: **heat lost equals heat gained**.

1.2 Heat transfer: conduction, convection and radiation

Heat is thermal energy that is in transit and it transfers (moves) from warmer matter to cooler matter, for example from the hot kettle to the cooler kitchen air.

There are three ways that thermal energy moves: by conduction, convection, and radiation.

Conduction

When heat is transferred by conduction the molecules within the material are agitated without any motion of the material as a whole.

The transfer of heat by conduction can be demonstrated if one end of a metal rod is heated to a high temperature, heat energy will be transferred down the metal rod toward the cooler end. This is because the higher speed particles from the hot end will collide with the slower ones, with a net transfer of energy to the cooler end of the rod.

Convection

Heat is transferred by convection by bulk motion of a gas or fluid such as air or water when heated. This causes movement away (see Figure 1.1) from the source of heat, carrying heat energy with it. Convection above a hot surface occurs because hot air expands, becomes less dense, and rises. Hot water likewise expands and is less dense than cold water and rises, causing convection currents which transport the energy.

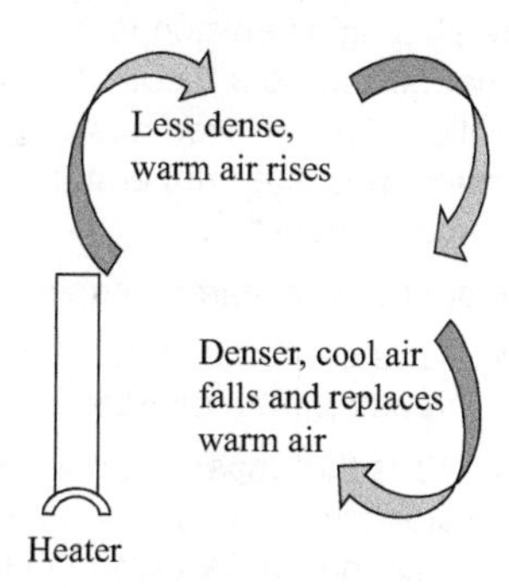

Figure 1.1 Convection

Radiation

Unlike conduction and convection, with heat transfer by radiation no particles are involved. All objects give out and take in thermal radiation, sometimes called infrared radiation. Infrared radiation is a type of electromagnetic radiation that involves waves and can transfer heat even through a vacuum. This explains why the Sun's heat can be felt even though it has travelled through space.

Some surfaces are better than others at reflecting or absorbing infrared radiation (see Table 1.1).

Table 1.1 Surfaces and thermal radiation

Colour	Finish	Thermal radiation emitting properties	Thermal radiation absorbing properties
dark	dull or matt	good	good
light	shiny	poor	poor

1.3 Hardening, annealing, normalising, case hardening, and tempering

Heat treatment is the process of heating a metal to a specific temperature, soaking it at that temperature and then cooling at a specified rate to change their physical and mechanical properties, without changing their shape.

There are three main stages to heat treatment:

Stage I Heating the metal slowly to achieve a uniform temperature throughout the material.

Stage 2 Soaking or holding the metal at the required temperature for a given time.

Stage 3 Cooling down the metal at an appropriate rate to room temperature.

Heat treatment of carbon steel

In its simplest form, steel is an alloy of iron and carbon (graphite). Figure 1.2 shows the relationship between temperature and carbon content of steel, and the transformation that occurs as a result of slow heating. At high temperatures, iron can take carbon into solid solution called **austenite**. On cooling, the austenite precipitates carbon to form **cementite**; the temperature at which this takes place varies with the percentage of carbon present.

From the iron–carbon equilibrium in Figure 1.2:

- A1 is the lower critical temperature and is the temperature at which austenite transforms to **pearlite**; austenite cannot exist below this temperature
- A3 is the **upper critical temperature**, and is the temperature below which ferrite starts to transform from austenite to a hypo-eutectoid alloy. Hypo-eutectoid steel has less than 0.8% carbon in its composition, and is composed of pearlite and ferrite.

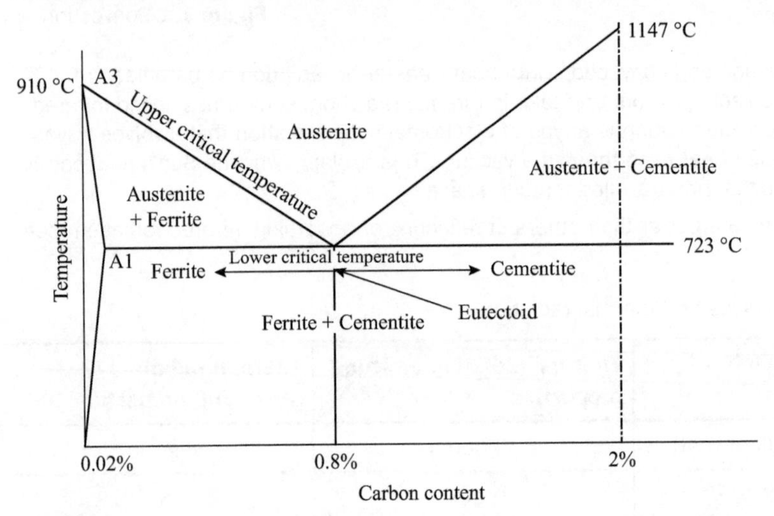

Figure 1.2 Carbon content of steel and temperature

Slow cooling will reduce the transformation temperatures, e.g. the temperature of the austenite boundary will reduce from 723 °C to 690 °C. Also, fast heating and cooling rates will have a significant influence on these temperatures, making the accurate use of Figure 1.2 difficult.

Key terms

- **Allotropy** – many metals can exist in more than one crystalline form. Above 906 °C iron changes from a body-centred cubic (BCC) structure (ferrite) to a face-centred cubic structure (austenite).
- **Austenite** is face-centred cubic (FCC) unit cell and is only possible in carbon steel at high temperature. It has a structure which can contain up to 2% carbon in chemical combination.
- **Carbon** strengthens steel and gives it the ability to be hardened by heat treatment. Carbon forms compounds with other elements called **carbides**, e.g. iron carbide, chrome carbide etc. It has a small atom size which fits into clusters of iron atoms.
- **Cementite** is very hard, but when mixed with soft ferrite layers its average hardness is reduced considerably.
- The **eutectoid point** is the lowest possible temperature of solidification for any mixture of specific constituents. In terms of an alloy of iron and carbon, it is the lowest melting/solidification point of any other alloy composed of the same constituents in different proportions.
- **Ferrite** has a body-centred cubic (BCC) structure which can hold very little carbon, typically 0.0001% at room temperature.

Hardening high carbon steel (>1.4% carbon) is the process of heating the steel to a temperature of 723 °C or above (austenite stage) and cooling it rapidly in oil or water to give the properties which enables it to resist plastic deformation, penetration, indentation, and scratching. In turn this will increase strength and improve wear-resistance properties. When steel is cooled rapidly from austenite, the FCC structure rapidly changes to BCC, leaving insufficient time for the carbon to form pearlite.

Pearlite is a mixture of alternate strips of ferrite and cementite in a single grain. A fully pearlite structure occurs at 0.8% carbon; further increases in carbon will cause cementite to form at the grain boundaries, which will weaken the steel.

Note: Mild and medium carbon steels (<1.4% carbon) do not have sufficient carbon to change their crystalline structure, and consequently cannot be hardened and tempered by heating and cooling. Medium carbon steel will become slightly tougher, although it cannot be hardened to the point where it cannot be filed.

A second form of hardening is **work hardening**, sometimes called strain hardening or cold working. This is the strengthening of a metal by plastic deformation such as by rolling, stamping, drawing, etc. An example of the impact of work hardening can be seen, on page 9 in this section.

Annealing is the process of slowly raising the temperature of the steel about 50 °C, above the austenitic temperature line. This temperature is held for sufficient time for all the material to transform into austenite or austenite–cementite. It is then very slowly cooled into the ferrite–cementite range. The steel is then cooled very slowly to room temperature. This produces a very soft structure with large grains, which have low toughness.

Normalising is a form of annealing, except after heating the steel to the recrystallisation stage it is cooled in air. This is more rapid than annealing and results in a structure that has a uniform (normal) grain size throughout. The steel is less ductile than that produced from annealing.

Tempering is the process of improving the characteristics of a metal, especially steel, by heating it to a high temperature but below the recrystallisation, then cooling it in air. This increases toughness, lessens brittleness, and reduces internal stresses. Temperatures for tempering vary considerably, depending on the type of steel and designed application; for tool steels, the hardness of which must be retained, the range is usually from 220 °C to 300 °C (see Table 1.2).

Table 1.2 Tempering colours

Colour	Degrees centigrade	Applications
pale straw	230	scrapers, lathe tools
dark straw	240	drills
brown	250	tin snips
purple	270	press tools
blue	300	springs

Examination of structures

There are two examination methods in metallography:

- macroscopy (macro)
- microscopy (micro).

Macro-examination of the structural or chemical characteristics of a metal is done by the naked eye, or with the aid of a low-power microscope (usually under 10 x). Specimens are ground, polished and etched to reveal the structure.

Micro-examination (by up to 1000 x) can provide information about:

- grain size
- grain boundaries.

Specimens are again prepared by polishing and etching, and then mounting for examination under a microscope (see Figure 1.3).

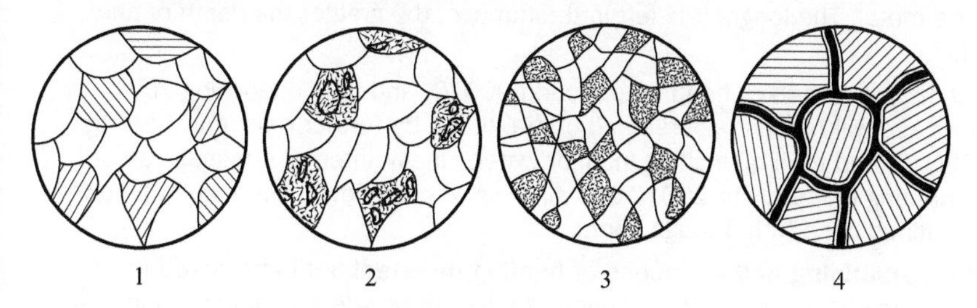

1 = Mixed grains of ferrite and pearlite
2 = Austenite transformed from pearlite
3 = Full Austenite transformation
4 = Ferrite will form at the grain boundaries, and a course pearlite will form inside the grains

Figure 1.3 Micro-examination of metals

1. Mixed grains of ferrite and pearlite; temperature is below 730 °C, therefore the micro-structure is not significantly affected.
2. Austenite is transformed from pearlite but the temperature is not sufficient to exceed the A3 line, therefore not all ferrite grains transform to austenite. Only the transformed grains will be normalised on cooling.
3. The temperature exceeds the A3 line a little, resulting in full austenite transformation. On cooling all grains will be normalised.
4. The temperature exceeds the A3 line significantly, permitting grains to grow. On cooling, ferrite will form at the grain boundaries, and a course pearlite will form inside the grains. A course grain structure is more readily hardened than a finer one, therefore if the cooling rate between 800 °C to 500 °C is rapid, a hard micro-structure will be formed.

Case hardening

Steels with less than 1.4% carbon cannot be hardened by heating and quenching.

Case hardening (sometimes called surface hardening) is the process in which a metal's surface is reinforced by adding a fine (thin) layer of another material. The case hardening of low carbon steel is by heating it above the austenite stage, around 910 °C, in a carbon-rich material. Being held at this temperature allows the surface to increase the percentage of carbon to a level where it can be successfully heat treated. In fact the resulting material may have properties more suitable for some applications than high carbon steel, because the surface can be hardened to improve its wear-resistance while retaining the characteristics of low carbon steel at its core.

There are four main methods of case hardening of low carbon steel:

- **pack carburising**
- **open hearth carburising**
- **cyaniding**
- **nitriding.**

With **pack carburising** the component is packed surrounded by a carbon-rich compound and placed in the furnace at 900 °C. Over a period of time, carbon will penetrate the surface of the metal. The longer it is left in the furnace, the greater the depth of hard carbon skin.

The material can be refined by heating to around 900 °C and then quenching in oil or water. The case will be hard, course and brittle. It can be further refined by heating to around 780 °C and then quenching in oil or water. Finally, internal stresses can be removed by heating uniformly to 200 °C and allowing to cool slowly. The steel will now have a hard refined case with a tough core.

Open hearth carburising is the process of heating the steel until cherry red in colour, and dipping it in carburising compound such as powdered charcoal and then reheating and re-dipping three or four times. Excess charcoal is cleaned off, and it is reheated and then hardened by quenching in water. The maximum depth of case hardening obtained by this method is 0.012 mm. Annealing and refining are not normally needed.

Cyaniding case hardening is fast and efficient. The pre-heated steel is dipped into a heated cyanide bath and allowed to soak. On removal, it is quenched and then rinsed to remove any residual cyanide. This process produces a thin, hard shell that is harder than the one produced by carburising and is completed in 20 to 30 minutes. A major drawback is that cyanide salts are a deadly poison. For this reason, the cyaniding material is kept under lock and key plus an appropriate antidote must be available.

Nitriding is the process in which the surface of alloy steels is hardened by heating in the presence of a compound of nitrogen and hydrogen (ammonia). Nitrogen released by the decomposition of the ammonia reacts with the metal to make iron nitride, a hardening substance. This means that the steel does not need to change from ferrite to austenite, and it retains its ferrite BCC structure.

Quenching

The rate of cooling is critical to achieving the required structure and the consequential properties of the material.

Quenching mediums

- **Air cooling** allows grain growth and is used for process such as annealing and normalising processes.
- **Brine** (water and salt solution) gives a quicker cooling velocity (rate) than water alone, giving a fine grain structure. There can be a possibility of internal stress and distortion of the material if the cooling is too rapid. Annealing may be needed to relieve stresses in the material.
- **Furnace cooling** is when the furnace is switched off and the material is left in to cool down very slowly. This is used for full annealing and gives maximum grain growth.
- **Oil** has a major advantage over water due to its higher boiling range, giving a slightly slower cooling rate than water which produces less stress in the material.
- **Water** is used to quench when a fine grain resulting in a hard material is needed.

- An alternative to furnace cooling is to quench the material in **sand** of varying temperatures; furnace cooling can be problematic because the equipment needs to be in use continuously. Hot sand (100 °C–200 °C) gives a slower cooling velocity (full annealing) than cooler (ambient) or damp sand, although the surface may need surplus sand removed on cooling.

Heat treatment of non-ferrous metals

Generally heat treatment of non-ferrous metals is annealing carried out to counteract the effects of work hardening caused by: drawing, rolling, stamping, forging, etc.

Aluminium is hardened by cold working and is annealed by heating to a temperature of 350 °C–400 °C and quenching in water, or air cooling.

Brass is hardened by cold working and is annealed by heating between 535 °C to 670 °C, and cooled in either air or water. Normalising is done by heating it up to 700 °C, and cooling either in air or water.

Copper tends to become hard and brittle when cold worked. It can be annealed by heating to a dull red colour (650 °C), soaking it at this temperature for a few minutes, and then quenching in water or allowing it to cool in the air.

Why don't you?

Explain the steps involved in the following heat treatments of low alloy (less than 0.2% carbon) steels. Discuss the expected microstructure for each treatment and how it affects the mechanical properties.

- Normalizing
- Quenching
- Tempering
- Annealing

Compare your notes with those of a classmate. Discuss what went well and any areas of difficulty. Where are the convergences and divergences between each of your findings?

Put a copy of your work in your portfolio of evidence.

1.4 Heat sources

Heat transfer methods

Heat can be transferred to the work by:

- **natural convection**
- **forced convection,** or by
- **radiant heat sources**.

Natural convection is fast, but not necessarily uniform, while forced convection is more uniform. Both methods are flexible, controllable and suitable for irregular shapes. Radiant heat transfer is faster at higher temperatures and relatively less expensive, but lacks the flexibility of convection and can distort irregular shapes. Additionally to the different types of transfer, ovens and furnaces can be designed to contain special atmospheres such as argon, carbon or nitrogen, to support heat treatment processes.

Flames, heaths, and ovens for heat treatment

Applied heat energy methods

The oxy-acetylene flame is made of two main components: acetylene is the fuel and oxygen is the oxidiser. Other pairings of fuel and oxidiser are used. Table 1.3 shows some common fuel and oxidiser pairings with their relative flame temperatures. Both fuel and oxide are delivered under pressure, which will affect the resulting working temperature of the flame significantly.

Table 1.3 Common fuel and oxidiser pairings

Fuel	Oxidiser	Temperature/°C
acetylene	air	2500
acetylene	oxygen	3470
butane	air	1980
gasoline	air	2140
propane	air	1990
propane	oxygen	2525

It is important that the fuel and oxidiser are in balance (see Figure 1.4). An **oxidising flame** is when there is excess oxygen, which in most cases is unwelcome in heat treatment. Where there is an excess of fuel, this is a **carburising** or reducing flame which will encourage the production of carbides. When the fuel and oxidiser are balanced, a **neutral** flame is produced.

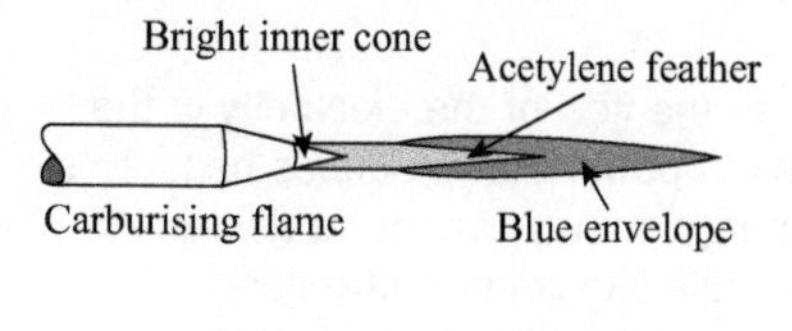

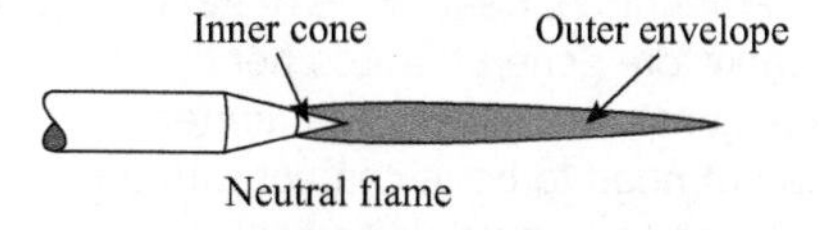

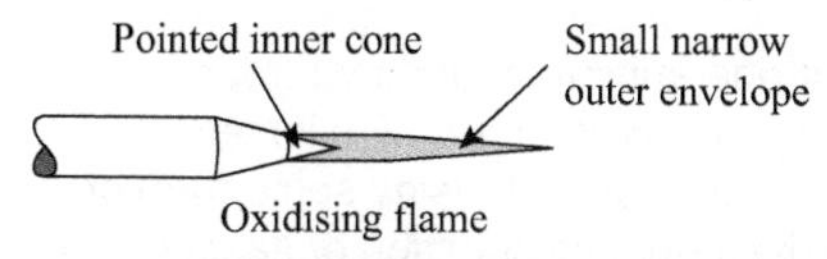

Figure 1.4 The balance of the fuel and oxidiser

An **open forge**, sometimes called a blacksmith's forge, uses coke, coal or charcoal as the fuel; the oxidiser is either hand bellows or a compressor (air). Like the acetylene torch, it is not easy to know precisely the temperature of the material so the colour is used to guide the forger.

Table 1.4 shows the approximate temperature for each colour of steel. The colours will vary for the equivalent temperatures for other materials.

Table 1.4 Approximate temperature for each steel

Colour of heated steel	Temperature/°C
bright yellow	1900
orange-yellow	980
orange-red	870
bright red	810
dull red	650
grey	425
blue	300
purple	270
brown	250
straw	230
light straw	215

Induction heating

While metals conduct electricity, they offer resistance to the flow of the electricity in the form of resistance to the flow of current causing a loss in power that produces heat. This is because, according to the law of conservation of energy, energy cannot be destroyed but only transformed – here into heat energy. Some metals like copper have good electrical conductivity and consequently generate low amounts of heat, while metals such as steel have lower electrical conductively and therefore generate more heat. When a coil carrying a current is wound around a steel component heat is generated and the steel component becomes hot. The coil does not need to be in contact with the component for the process to work. This application is used in induction hardening.

Ovens

Ovens are generally classified as heating equipment operating from ambient (room temperature) to 538 °C, while furnaces operate to higher temperatures. Equipment may be designed for batch loading, or for a continuous flow of work using some form of conveyor system. The source of heat is normally derived from combustion of gas, oil, electricity, hot water or steam

Selecting equipment

There are many issues that must be considered when selecting industrial heat treatment equipment, including:

- the quantity of components and their material
- the complexity, uniformity, size and shape of the component
- the permissible tolerances in temperature variation
- whether batch or continuous processing is required.

Batch ovens are the largest range of ovens used to manufacture products and are classified as laboratory, cabinet or truck-in. The chamber size can range from small bench top units to large industrial installations with hundreds of cubic metres.

Continuous ovens have construction features that are similar to batch ovens. The difference being continuous ovens include features such as the means of conveying components and continuous product loading methods.

Remember: The three phases of heat treatment are:

- heating-up
- soaking, and
- cooling-down.

These are critical elements to consider for selecting the most appropriate heat treatment equipment. There are five basic questions to ask:

1. Does the oven have sufficient heating capacity to bring the components to the required temperature uniformly within the specified time?
2. Must the heat-up rate be at a controlled one and maybe staged, or can the component be allowed to reach temperature as quickly as possible, given the oven's heating capacity?

3. Is the atmosphere in the oven to be controlled; is it an inert gas or other atmosphere?
4. How accurately must the soaking temperature and time be controlled?
5. Is the cooling down to be controlled by exhausting the oven with ambient or cooled air?

Why don't you?

1. Straighten out two metal paperclips using pliers. Bend one through 90° repeatedly until it breaks. Count how many times the paperclip was bent until it fractured. Take the second and bend it half as many times as the number that resulted in the failure of the first. Now heat the second paperclip until bright red, hold it at that temperature for several minutes and then allow it to cool down slowly. When it is at room temperature bend it through 90° until it breaks, noting how many times it bent until it fractured. Add the first number to the second after annealing.
   - What was the cause of both failure of both paperclips?
   - Why could the second paperclip still withstand the same number of bends, even though it had been subjected to bending previously?
   - Explain the heat treatment process carried out and how this impacted on the physical properties of the material.
2. **Safety: Ask your teacher's permission and wear eye protection. Take care with hot metals.**

   If you have access to oxy-acetylene equipment, try setting the flame to:
   - a neutral flame
   - an oxidising flame
   - a carburising flame – apply this flame to a clean piece of metal. Why does soot form on the metal?

   Draw and label each of the flames.

   Put a copy of your work in your portfolio of evidence.

2.1 Single, double, and multi-point cutting tool theory

There are three main cutting tools: single, double and multi-point.

- **Single point** – these cutting tools are usually wedge-shaped, such as lathe and boring tools. Usually the workpiece rotates relative to the cutting tool.
- **Double point** – these are often fluted tools such as twist drills and reamers. Normally the cutting tool rotates relative to the workpiece.
- **Multi-point** – these include milling cutters and grinding wheels. Again the cutting tool rotates relative to the workpiece, while the workpiece is moved from front to back and/or side to side.

Cutting tool material should be able to:

- resist wear at high temperatures by having high hardness
- reduce cutting edge temperature by having high thermal conductivity
- resist cutting edge deformation under chip formation stresses
- resist cutting edge fatigue due to intermittent cutting
- maintain sufficient rigidity to achieve accuracy of cutting action
- resist adhesion (lubricity) when cutting soft, ductile material.

The main angles on the standard knife tool are shown in Figure 2.1

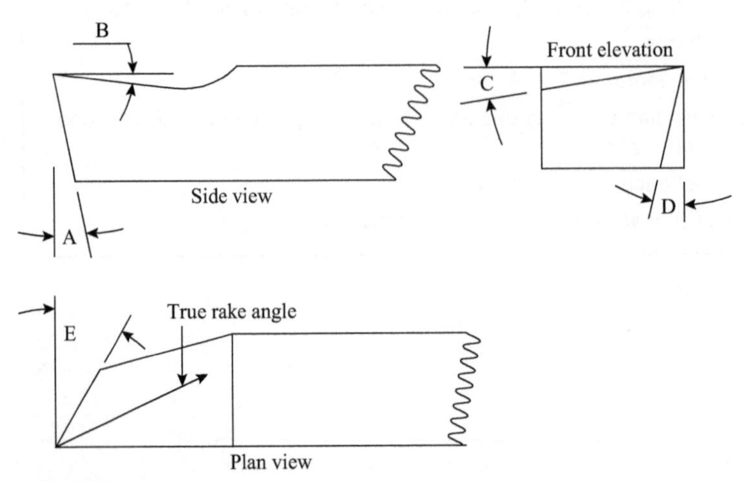

Figure 2.1 A standard knife tool

A Front clearance angle

B Top rake angle

C Side rake angle

D Side clearance

E Trailing clearance

The tool wedge angle is the body of the cutter between the rake and clearance angles.

The rake angle varies most and generally increases as the materials get softer, more ductile. The front clearance angle varies least, and should be no more than is necessary to avoid contact with the workpiece. Excessive clearance will make the tool point too sharp, losing strength, and heat dissipating volume. Boring tools generally require more clearance because of the shape of the workpiece. Rake angles "B" and "C" combine to make the true rake angle. An appropriate lubricant and coolant will improve cutting performance and increase tool life. Angles "A" and "D" should be as small as possible, which will improve heat dissipation and tool strength and consequently increase tool-life.

Figure 2.2 shows a double (point) fluted twist drill:

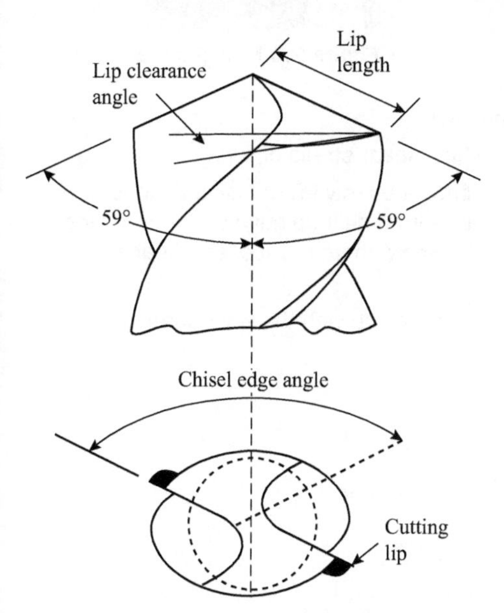

Figure 2.2 Double (point) fluted twist drill

- The cutting lips must be equal length and at an angle of 118° (59° + 59°), the most common angle used and suitable for mild steel. Stainless steel performs better with an angle of 135°, and 120° for toughened steel. If either or both lip length angles are unequal, the hole produced will be oversized and the drill life will be shortened.
- The lip clearance must begin immediately after the cutting edge and be no more than 12° to maintain strength and maximum heat dispersal.
- The chisel edge angle formed between the chisel edge and the cutting lip should be 125° to 135°.
- For drilling thin material, a cutting lip angle of 140° is used.

- The face milling cutter in Figure 2.3 cuts on both the circumference, and part of the peripheral edge.
- The rake angle is formed by the centre line of the cutter to the cutting edge, and the angle face of the tooth. The rake angle can be positive, zero, or negative, depending on the material being cut.

There are two types of clearance angles on a milling cutter: the primary clearance angle, and the secondary clearance angle.

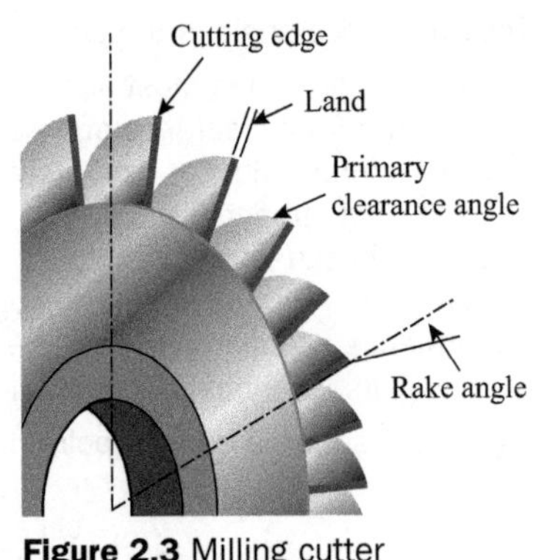

Figure 2.3 Milling cutter

See Figure 2.4 for single point cutting tool terminology.

- The rake angle is the angle formed by a line perpendicular to the workpiece and the tool face.
- The shear plane is the plane in the line at which shear stress occurs.
- The chip is the metal removed and can be continuous (**swarf**) or discontinuous (**segmental**). A third type of chip is continuous with a built up edge on the tool face, which occurs when machining ductile material where there is excessive heat and pressure.
- The shear angle is the angle formed by the shear plane with the work surface.

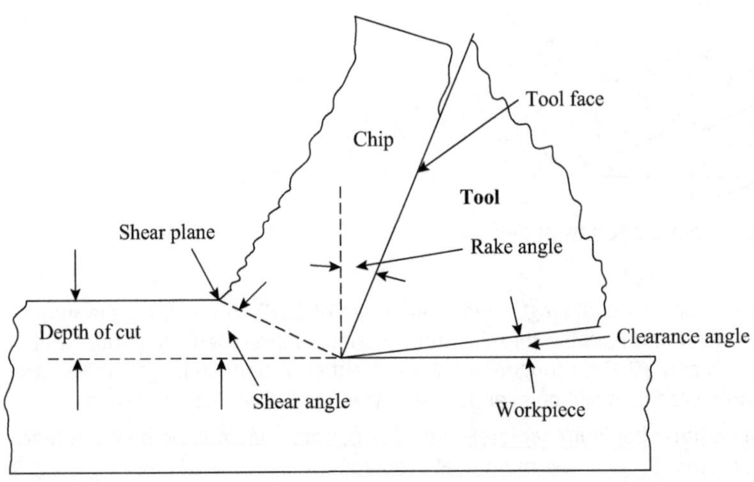

Figure 2.4 Single point cutting tool terminology

Orthogonal cutting is where the cutting edge of the tool is set to cut perpendicularly to the direction of motion producing a smaller shear cutting force area.

Oblique cutting is where the cutting edge of the tool is set at an acute angle to the direction of motion producing a larger shear cutting force area. Helical shaped milling cutters will have an oblique cutting action.

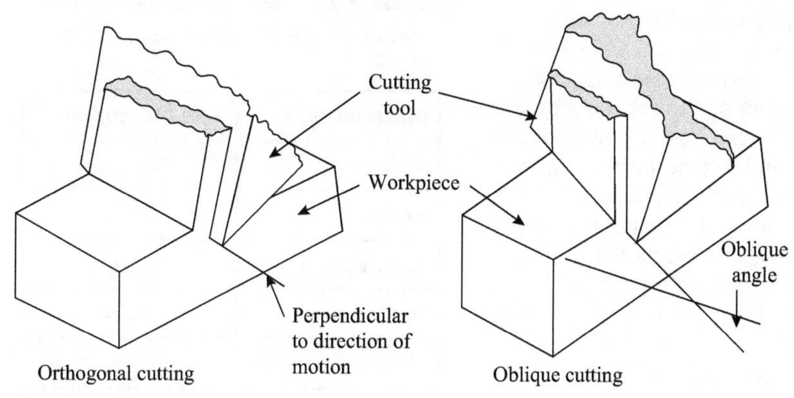

Figure 2.5 Cutting mechanisms

2.2 Factors that affect the life of a cutting tool and tool geometry

Tool life is affected by many factors such as: metal remove rate (cutting speed, feed rate, and depth of cut), tool geometry, cutting material and metal being cut, the cutting fluid, and rigidity of the machine.

1. Metal removal rate (MRR) is the combination of cutting speed, feed rate and depth of cut. It has a profound impact, individually and collectively, on tool life. As the MRR increases, the temperature rises. The heat is more concentrated in the cutting tool than in the workpiece, and the hardness of the cutting tool can be changed. It is important to realise there is a strong link between MRR and productivity. The formula for calculating MRR is given on page 20 of this section.
2. Tool geometry directly affects tool life as it defines the magnitude and direction of the cutting force, and the distribution of the thermal energy released in the cutting process. Tool geometry also defines the direction of chip flow along with controlling chip breakage and evacuation. The geometry will only be appropriate if the lathe tool is set correctly on tool height.
3. Generally speaking the cutting tool must be at least $1\frac{1}{2}$ times harder and tougher than the material being cut, it must resist wear and have higher thermal stability. Selecting the right cutting material for the workpiece is central to maximising tool life.
4. Cutting fluids and oils reduce friction, and dissipate heat generated between workpiece and tool during the machining process. The correct coolant will also aid chip removal from the cutting zone.

5. Machine instability may arise from poorly mounted foundations or dampers, or poor conditions of the machine from excessive wear due to poor maintenance. The resulting vibration leads to machine and/or tool chatter. This will cause shock loading of the cutting tool and may produce tool fatigue and failure.

Metal removal rate

The cutting speed (V), at which the cutting tool travels over the workpiece, is expressed in m/min. It is important to use the correct speed for the material being cut, if it is too high it may cause a built-up edge and reduce tool life; if too low time is lost, and the production rates are low. Calculating the spindle speed (N) in rpm is set by dividing the cutting speed (V) by the circumference (πD) of the workpiece (turning) or diameter of the cutter (milling).

Here is the equation for calculating the spindle speed N, where N is in rpm; V is the cutting speed in m/min; D is the diameter of the workpiece in millimetres:

$$N = \frac{1000\,V}{\pi D} \approx \frac{300\,V}{D}$$

Table 2.1 Linear cutting speed for various materials when using HSS cutter

Material type	Speed/m/min
Aluminium	75-105
Brass	45-60
Bronzes	24-45
Cast iron	18-24
Mild steel	30-38
Steel (tough)	15-18

Worked example 1

Calculate the spindle speed of cutting mild steel of diameter 50 mm, at linear cutting speed of 35 m/min.

$$N = \frac{1000 \times 35}{\pi \times 50} \approx \frac{300 \times 35}{50} = 210 \text{ rpm}$$

Did you know?

The symbol $\approx$ means approximately. Using 3 for π is accurate enough for most situations.

It is important to ensure the cutting speed (V) and diameter of work are in the same units – multiply m/min by 1000 to convert to mm/min.

Rearranging cutting speed formulae

It may be necessary from time to time to calculate the diameter from a known rpm and linear cutting speed.

Table 2.2 The spindle and cutting speed formula for a milling machine

Choosing the cutter diameter		
• Start with the equation for the *N* rpm, a linear cutting speed of *V* (m/min) with a *D* cutter diameter (cm).	$N = 100V \div \pi D$	• Multiply both sides by π.
	$N\pi = 100V \div D$	• Divide both sides by 100*V*.
• End with an equation for *D*, the cutter diameter.	$N\pi \div 100V = D$	• Swap the equations around.
	$D = N\pi \div 100V$	

Feed rate

Feed rate is the distance the tool advances for each revolution of the workpiece (turning) or cutter (milling) The feed rate used depends upon a variety of factors, including power and rigidity of the machine, spindle horse power, depth and width of cut, design and type of cutting tool, and the material being cut.

Note: Feed rate/tooth is feed rate ÷ number of teeth on the milling cutter

There are two main types of feed rate:

1. **Roughing cut** is used to maximise economy of machining.
2. **Finishing cut** is used to produce the final surface finish needed.

Table 2.3 Some average feed rates for common material for rough and finishing cuts

Feed rates Material	Rough cuts/ mm	Finish cuts/mm
Mild steel	0.25–0.5	0.07–0.25
Tool steel	0.25–0.5	0.07–0.25
Cast iron	0.4–0.65	0.13–0.3
Bronze	0.4–0.65	0.07–0.25
Aluminum	0.4–0.75	0.13–0.25

Depth of cut

Depth of cut is half the reduction in the diameter of the workpiece (turning) during one revolution of the workpiece. The depth of cut will depend, as the feed rate does, on the power and rigidity of the machine, the design and type of cutting tool, and the material being cut. The aim is to use a few roughing cuts as possible and only one finishing cut. The depth of cut on a milling machine will be the distance the cutter penetrates the surface of the workpiece (see Figure 2.6).

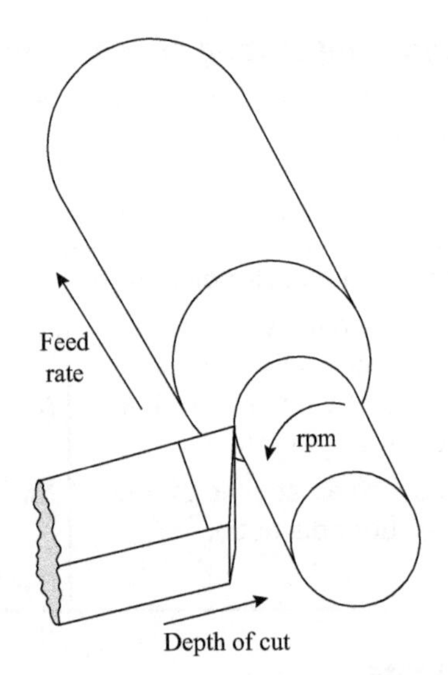

Figure 2.6 Feed rate and depth of cut

Calculating metal removal rate

Metal removal rate MMR (or Q) = rpm × dc × f, where f = feed rate, and depth of cut (dc) = difference between r_1 and r_2.

Worked example 2

Find the metal removal rate of mild steel with 210 rpm, $dc = 0.25$ mm, and $f = 0.5$ mm/rev rough cut.

$Q = \text{rpm} \times dc \times f = \text{mm}^3/\text{min}$

$Q = 210 \times 0.25 \times 0.5$

$= 210 \times 0.5 \times 0.5 = 26.25\,\text{mm}^3/\text{min}$

Note: Units of MMR are mm³/min.

2.3 Techniques used in sharpening tools

Before sharpening lathe tools it is important to be familiar with the name, shape and use of common lathes. If each tool is to do its job it must have the correct shape and angles. The tool shapes illustrated in Figure 2.7 will enable nearly all turning operations to be carried out. The external profile would be completed first supported by a tailstock centre. However, while machining the internal diameters of at least one fixed steady will be needed to support the work. The centre would be drilled to allow initial boring using a roughing tool, then finished using a finishing tool.

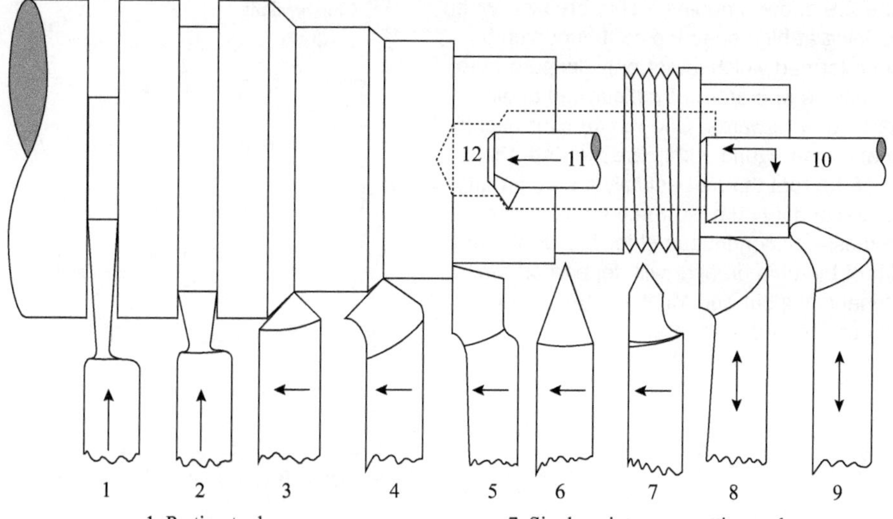

1. Parting tool
2. Recessing tool
3. Heavy duty roughing tool
4. 45° roughing tool
5. Right hand knife tool
6. Straight round nose (bull nosed) tool
7. Single point screw cutting tool
8. Facing tool
9. Cranked (bent) nosed tool
10. Finishing boring tool
11. Roughing boring tool
12. Drilled hole to allow boring

Figure 2.7 Common lathe tools

Most lathe tools are maintained by being re-sharpened on an off-hand grinder. Figure 2.8 shows the front and side clearances being re-sharpened on an off-hand grinder, with the tool rest adjusted to the appropriate angle. These types of grinders do not usually have coolant flowing. However, the operator can have coolant handy in a small bath so the tool can be cooled and lubricated from time to time.

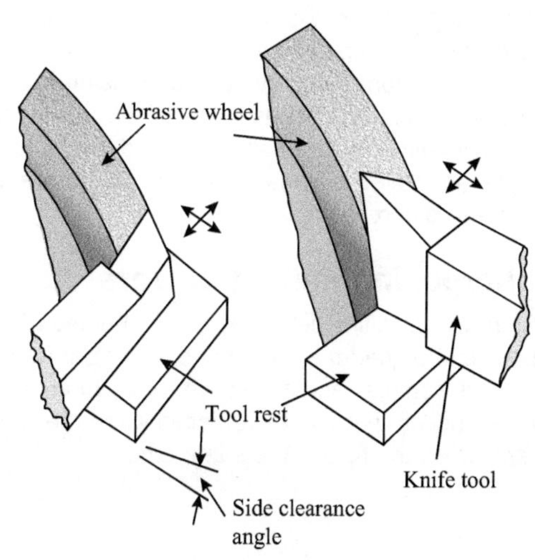

Figure 2.8 An off-hand grinder

Figure 2.9 shows grinding a chip breaker. When machining at high speed, a continuous chip may be formed which is not only dangerous to the machine operator but difficult to handle and dispose of afterwards. One way to control chip formation is to grind a chip breaker into the top face of the tool that is basically a step ground in, or secondary rake, so the continuous chip becomes a discontinuous chip. The depth and width of the chip breaker will depend on the material being cut and MRR – cutting speed, depth of cut, and feed rate.

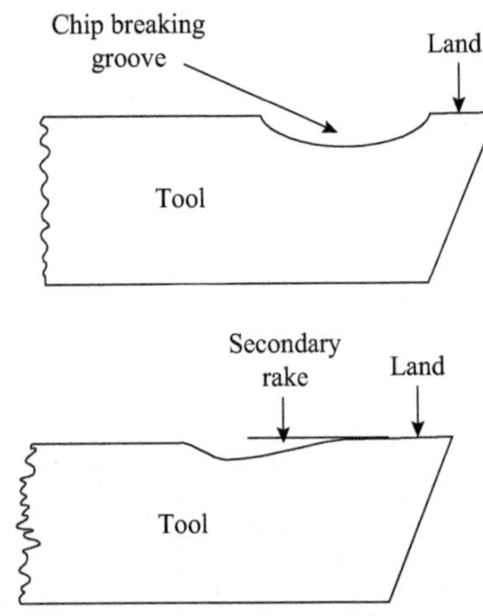

Figure 2.9 Grinding a chip breaker

Tool and cutter grinders

There are a wide variety of specialist lathe tool grinding machines that have manual or tool holding devices, to accurately produce angles and profiles for more precision turning than can be achieved with a tool ground on an off-hand grinder. Generally they have a cupped grinding wheel that allows more intricate shapes to be produced with precise angles. If a forming tool is to be used, for example a screw cutting tool, a gauge is used to ensure the correct profile for the tool is ground.

Sharpening drills

If grinding by hand it is important to, assuming the drill was correctly ground previously, to follow the shape and length of the lip. A light touch is needed to take the clearance angle back from the original lip and maintain its equal length. A handy tip is to put two hexagonal bolt heads together – they will form a 120° angle which is sufficient for hand grinding, while presented to the bolt heads for equality of lip angle and length. Again specialist drill sharpening machines are available that are adjustable for various diameters and lip angles.

Sharpening milling cutters

Sharpening milling cutters requires the use of a tool and cutter grinder, as it is not possible to grind them freehand. Milling cutters are relatively expensive, and re-grinding them is complex and expensive. Many modern milling cutters have tipped throwaway insert edges that can be replaced when they become dull. It is normally necessary to replace all the tips at the same time.

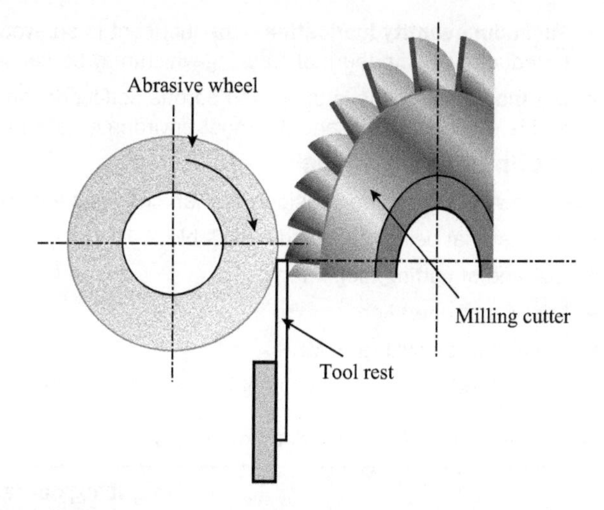

Figure 2.10 Tool and cutter grinding of miller cutter

The land and primary clearance of milling cutters with straight teeth can be re-ground with a tool and cutter grinder as shown in Figure 2.10. The tooth support ensures each tooth is ground to the same angle and the circularity of the milling cutter maintained. In practice, the grinding wheel revolves downward toward the cutting edge, so that the action of the wheel forces the tooth against the tooth rest. Any burrs can be removed with a hand-held oil stone.

2.4 Cutting fluids and their uses

Cutting fluids are used to:

- reduce friction, tool wear, and consequently increase tool life
- dissipate heat from the cutting tool and cutting area
- reduce cutting forces and energy consumption
- carry away chips from the cutting area
- offer corrosion protection (some lubricants).

Types of cutting fluid

There are four general types of cutting fluid:

- **Straight oils** – animal, mineral, synthetic oils, and vegetable. They are effective lubricants but less effective coolants.
- **Emulsions** – a soluble mixture of oil and water, sometimes with additives. They are effective coolants but less effective lubricants.
- **Semi-synthetics** – chemical emulsions containing a little mineral oil. Synthetic fluids can provide the best cooling performance among all cutting fluids.
- **Synthetics** – chemicals with additives. The cost and heat transfer performance of semi-synthetic fluids lie between those of synthetic and soluble oil fluids.

Applying cutting fluids

There are four main methods of applying cutting fluids:

- **Continuous flooding** – this is the most common delivery method.
- **Spray mist** – this is a coolant combined with a stream of high pressure air.

- **Minimum quantity lubrication** – the lubricant is sprayed directly onto the cutting tool to reduce possible thermal cracking which may be caused by coolants.
- **Dry machining** – this is used with porous self-lubricating material such as cast iron, and is the least costly and the most environmentally friendly.

Selecting cutting fluids

The following need to be considered when choosing a cutting fluid:

- the material being machined (see Table 2.4)
- the type of cutting tool being used
- the type of operation
- the metal removal rate, MRR
- the method used for the coolant delivery to the cutting area.

Table 2.4 Cutting fluid for various materials

Material	Machining operation		
	Drilling	**Turning**	**Milling**
Aluminium	soluble oil or lard oil	soluble oil	soluble oil, lard oil or synthetic oil
Brass	dry, soluble oil or lard oil	soluble oil	soluble oil or dry
Cast iron	dry or soluble oil	soluble oil or dry	dry or soluble oil
Copper	dry, soluble or synthetic oil	soluble oil	soluble oil or dry
Steel	soluble or synthetic oil	soluble oil	soluble or synthetic oil

Why don't you?

1. Research what a slitting saw milling cutter is, and what it is used for. Draw it in profile showing the geometry of the teeth.
2. A 60 mm diameter piece of brass is to be turned using a tool steel cutter. Calculate the cutting speed in revs per minute. Use the linear cutting speeds from Table 2.1, on page 18.
3. While machining the piece of brass in question 2 a feed rate of 0.4/rev is used with a depth of cut of 0.5 mm. Calculate the MRR for this operation.
4. Suggest what type of cutting fluid (if any) should be used for question 3, and suggest an appropriate application method if one is used.
5. Which type of cutting fluid might offer corrosive protection?

Share your work with a classmate. Say what went well and what could have been carried out more effectively. Put a copy of your work in your portfolio of evidence.

3. Reading and interpreting engineering drawings

Option B Section 2.5

3.1 Scale engineering drawings

Engineering drawings are rarely drawn full (life) size, which is a scale of 1:1. This is because of the impracticalities of drawing large items such as vehicles, doors, etc. Drawings are needed that show an object dimensioned with actual (physical) sizes, but reduced (or enlarged) by a certain degree. For example, if a drawing has a scale of 1:10, anything drawn length 10 mm would represent an actual size of 100 mm, etc.

Table 3.1 Drawing and measuring scales

Drawing scale	Measuring scale	Drawing scale	Measuring scale
1:20	1 cm = 0.2 m	1:500	1 cm = 5 m
1:25	1 cm = 0.25 m	1:1000	1 cm = 10 m
1:50	1 cm = 0.5 m	1:1250	1 cm = 12.5 m
1:100	1 cm = 1 m	1:2500	1 cm = 25 m
1:200	1 cm = 2 m	1:5000	1 cm = 50 m

Full size	1:1	Life size
Reduction	1:2	Half size
	1:5	Fifth full size
	1:10	Tenth full size
	1:20 and so on	One-twentieth full size
Enlargement	2:1	Twice full size
	5:1	Five times full size

Note: The drawing paper used should be big enough to allow the largest scale as possible, while leaving space for headings, symbols, abbreviations and notes. The scale used must be shown in the **title block** on the drawing.

Representing scales: The ratio between the drawing and the actual size of the component can be represented in one of two ways:

- **Scale** – for example, 1 cm = 1 m or 1 cm = 100 cm or 1:100
- **Representative fraction (RF)** – this is the scale expressed as a fraction, for example $\frac{1}{100}$, i.e. the ratio between the sizes of the component and the drawing.

3.2 General principles of dimensioning

Here are some principles of dimensioning:

- **Dimensions** are the numerical values expressed in appropriate, usually metric, units of measurement. They are shown on technical drawings with lines, symbols and notes.
- The **outline** and **profile** of components is shown as a thick continuous line.
- Small gaps are left between feature's dimension and extension lines.
- Dimension lines have arrowheads at both ends that are filled in (see Figure 3.1).

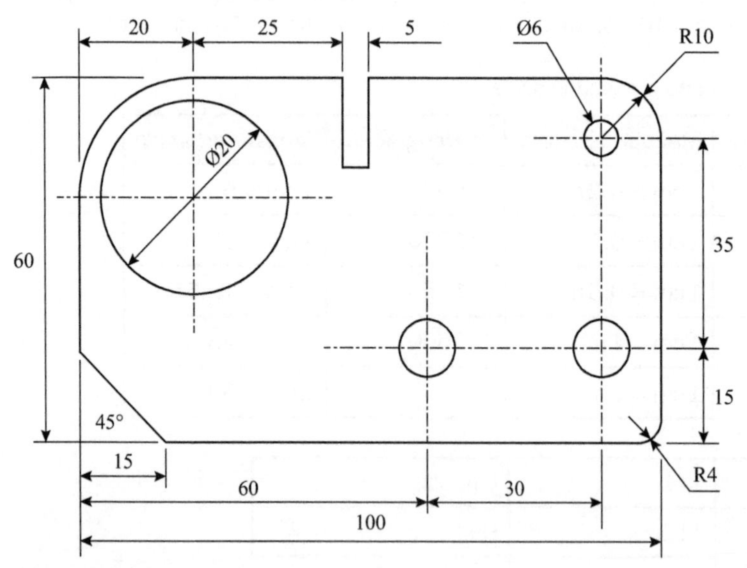

Figure 3.1 Technical drawing

- Horizontal and vertical axes of holes are shown with a centre line extending out to be dimensioned.
- Extension lines project a little beyond the dimension line.
- Dimension lines should be positioned outside the profile of the component.
- Dimensions are written above horizontal dimension lines and above the vertical dimension lines when viewed from the right-hand side of the drawing.
- Dimensions are in millimetres (symbol mm), unless otherwise stated, so it may be omitted.
- The symbols ∅ = **diameter** and ***R*** = **radius** are positioned before the dimension (see Figure 3.2).

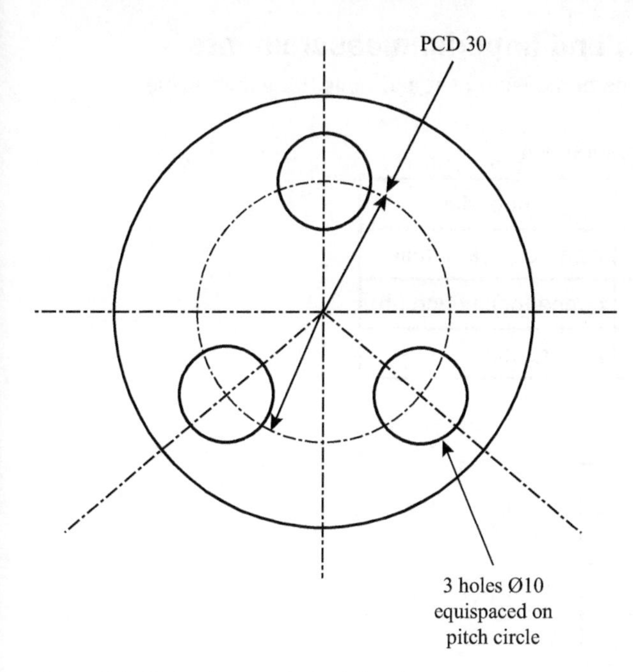

Figure 3.2 Holes in a technical drawing

- Where there is symmetry of features on a centre line it is permissible to dimension only one half.
- Non-functional dimensions are denoted by the abbreviation **NF**.
- Dimensions should never be calculated from the other dimensions shown on the drawing.
- Dimensions should be expressed to the least number of significant figures, e.g. 120 **not** 120,0.
- The decimal marker shall be a bold comma.
- Zero should precede a decimal of less than one, e.g. 0**,**25.
- Angular dimensions shall be expressed in degrees and minutes, e.g. 45° and 27° 30', or 27,5°.
- Preferred material stock sizes and standard components, such as sizes of nuts and bolts, should be used whenever possible.
- All dimensions, except auxiliary ones, are subject to general tolerances of ±0.4 linear ±0° 30′ or ±0**,**5°, unless otherwise stated.
- See section on tolerances, Section 5.3, page 41.
- Notes can be used to give greater clarity if needed.

3.3 Converting metric and imperial measurements

Table 3.2 gives the common units of measurement and Table 3.3 shows some conversion factors.

Table 3.2 Common units of measurement

Aspect	Metric	Imperial
Length	mm, cm, m, km	inch, foot, yard, mile
Mass	mg, g, kg	ounce (oz), pound (lb)
Capacity	ml, cl, l	pint, gallon

Table 3.3 Converting units

Metric and imperial
1 m = 39.37″
1′ = 30.5 cm
1″ = 2.54 cm or 1″ = 25.4 mm
1 kg = 2.2 lb
1 gallon = 4.5 litres
1 litre = $1\frac{3}{4}$ pints

Worked examples

Change 10′ into mm.
25.4 × 10 = 254 mm

Change 44 lb into kg
44 × (1 ÷ 2.2) = 20 kg

Change 9 litres into gallons
9 ÷ 4.5 = 2 gallons

Change 0.02″ into metric
0.02 × 25.4 = 0.508 mm or 0.0508 cm

Change 0.06 mm into imperial
0.06 × 0.03937 = 0.00236″

Temperature

Water boils at 100 °C and at 212 °F, and freezes at 0 °C and 32 °F.

The difference between 212° and 32° is 180°.

To convert degrees Celsius, °C , to degrees Fahrenheit, °F:

$$°C = (°F - 32) \times \overset{5}{\cancel{100}} \div \overset{9}{\cancel{180}} = °C = (°F - 32) \times 5 \div 9$$

$$557\,°F \text{ as } °C = (557 - 32) \times 5 \div 9 = 292\,°C$$

Work and energy

Converting foot pounds (ft–lb) to newton-metres (N m):

A foot–pound is the unit of work and is equal to the work done by a force of one pound acting through a distance of one foot in the direction of the force.

A joule is the SI unit of work or energy, and is equal to the work done by a force of one newton when its point of application moves one metre in the direction of action of the force.

$$1\text{ ft–lb} = 1.356\text{ N m, and } 1\text{ N m} = 0.737\text{ ft–lb}$$

$$\text{So, } 12\text{ ft–lb} = 16272\text{ N m and } 21\text{ N m} = 15.5\text{ ft–lb}$$

Note: 1 N m = 1 J

Why don't you?

Draw the component in Figure 3.3 on a scale of 1 : 25. Start by drawing a title block including the scale.

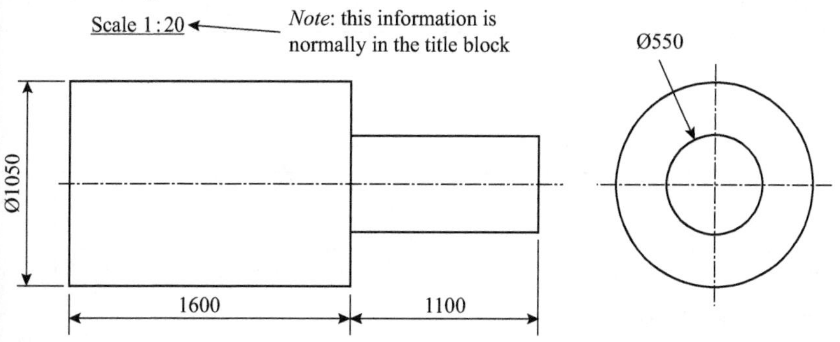

Figure 3.3 Technical drawing

Starting point

- Where will you get a 1 : 25 scale rule?
- What is the most appropriate size of paper to use for the drawing?

Work with a classmate and check each other's work. What was done well and what could have been improved?

Reflection

1. Give at least two examples of where you might find scaling used in everyday life.
2. What professions do you think need to be able to work with scaling?
3. How important is it to still include tolerances on scaled drawings?

Sign and date your work, and put a copy in your portfolio of evidence.

4. Preparing to produce technical drawings

Option B Section 2.4c

4.1 Producing drawing layouts

Drawing layouts for a title block

ISO 7200 Technical drawings

The title block identifies the requirements to be used on engineering drawings. It is normally placed in the bottom right of the drawing frame, and it should contain the following information:

- the owner or name of the company
- the title of the drawing
- the unique drawing number
- the scale used
- the angle of projection used
- the name/initials/signatures those involved along with appropriate dates
- plus other specific information.

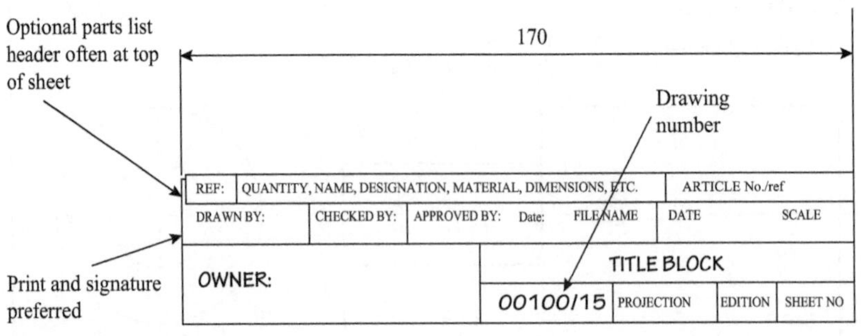

Figure 4.1 The title block

Drawing layouts for a parts list

If the drawing contains a number of parts, a table of parts list is attached to the bottom right just above the title block (see Table 4.1).

The **parts list** normally has the following information:

- each part's ID number
- the name part
- the drawing number of each part
- the quantity needed
- material specifications.

Table 4.1 A parts list

PART NO.	PART NAME	DRAWING NO.	QTY.	MATERIAL
PARTS LIST				

Drawing layouts for a revision list

Normally a **revision table** is located in the upper right of the drawing and sometimes at the bottom on the left side of the title block (see Table 4.2). It records all changes resulting from modifications to the drawing, such as:

- the number of the revision
- a description of the revision
- the date of the revision
- who approved the revision.

Table 4.2 A revision list

REVISIONS			
REVISION NO	DESCRIPTION	DATE	APPROVED

4.2 Lettering, line types, and quality

Types of line

Technical drawing lines are used to communicate specific information for designers, engineers, manufacturers, etc. when reading drawings. These line types are a technical language that convey important technical information. It is essential that you refer to these conventions when producing drawings (see Table 4.3).

Table 4.3 Types of line used in technical drawings

Type of line	Description	Example
Outlines and surroundings of the parts and components	Continuous thick line	________________
Measure lines, intersection, leader, hatching and section lines	Continuous thin line	________________
Axis of components	Centreline	_ ___ _ ___ _ _
Hidden detail	Dashed thin lines	- - - - - - - - - - - - -
Dimension lines	Continuous thin line with arrows	◄────────────►
Lines of symmetry and section planes	Dashed thin lines with dots	—·—·—·—·—·—·-
Cutting plane – arrows showing sectional view	Thick dashed line with arrows	↓- - - - - - - - - - -↓
Position line	Thin dashed line with two dots	—-- --—-- --—-- ---
Surfaces to be processed	Dashed thick lines with dots	—·—·—·—·—·—·—

Lettering

The best pencils for general lettering are H, F and HB grades. Many draughtsmen prefer a chisel point when lettering as this gives the opportunity to vary the thickness of the stroke, which is also true when drawing lines. Figure 4.2 shows the same chisel point pencil viewed from two angles.

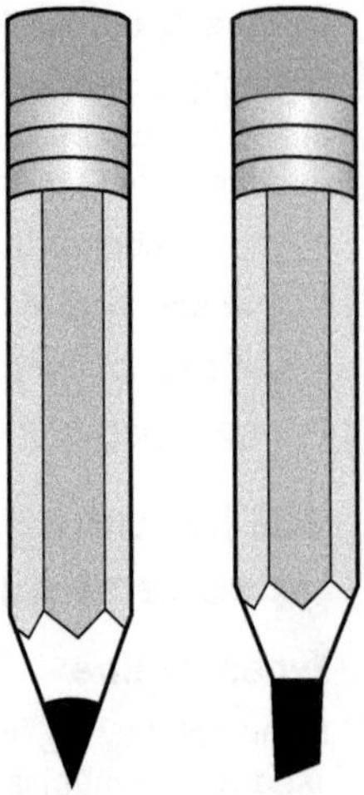

Lettering is a writing skill and needs to be practised (see 'Why don't you' below). It is important to produce lettering where there are proportions of height and width. Letters are normally slightly taller than their width, with the exception of "M" and "W" which are as wide as they are high. Letter spacing needs to be even and consistent. The pencil is best held at about a 60° angle to the page. The convention is to use only capital letters on drawings to ensure clarity and consistency.

Figure 4.2 A chisel point pencil

The right sequence of strokes will help to form letters that are legible and in proportion. In Figure 4.3 the arrows at the side of the letters indicate sequence and direction. The vertical strokes are done first and then the horizontal, with the exception of the "T" where the top is drawn first and then the vertical. Lines and circles are produced as shown in Figure 4.4.

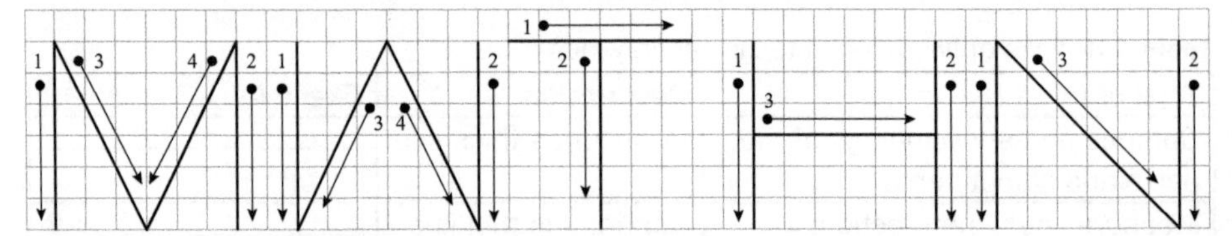

Figure 4.3 Drawing straight line letters

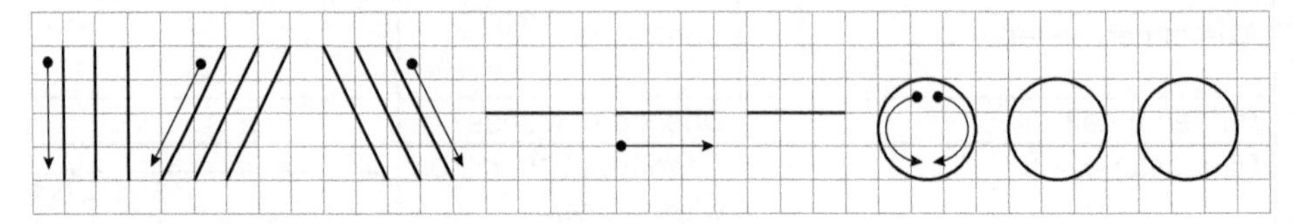

Figure 4.4 Drawing lines and circles

Gothic vertical lettering (see Figure 4.5) is the most commonly used font on drawings, although an inclined to the right version is also acceptable.

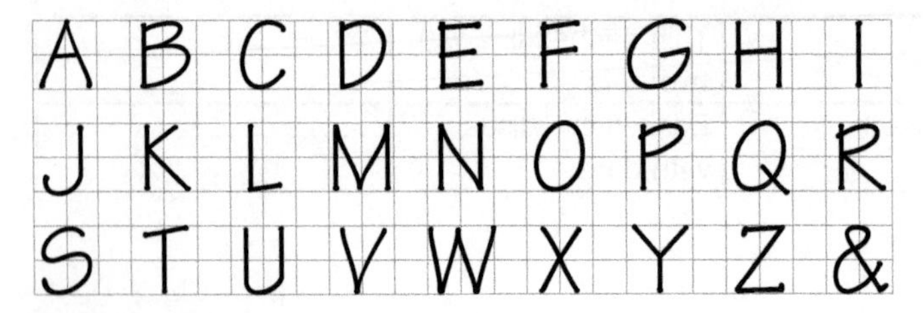

Figure 4.5 Gothic vertical lettering

Numbers are drawn in the same way and to the same proportions as shown in Figure 4.6.

Figure 4.6 Drawing numbers

Fractions (see Figure 4.7) are drawn twice as tall so that the numbers of the fraction element is the same size as the integer element. However, remember a whole number with a fraction is *not* an integer.

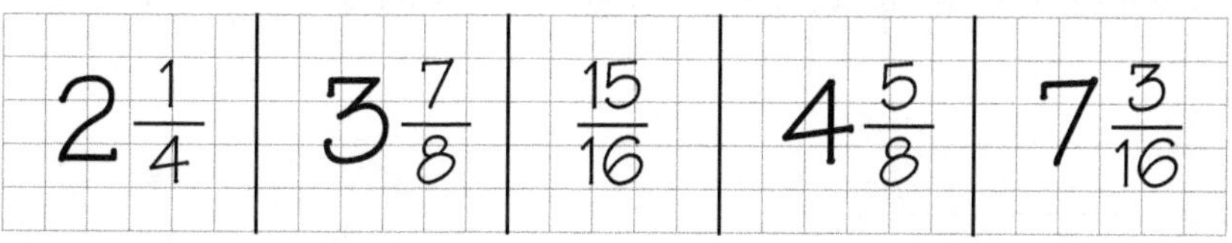

Figure 4.7 Drawing fractions

Note: The line between the numerator and denominator of the fraction is always horizontal, it is never at an angle and is only as wide as the numbers.

Why don't you?

Practise, practise, and practise!

Some people are naturally better at lettering than others, but all of us can improve with practice and more practice.

Guidelines are helpful. Using graph or squared paper helps to achieve uniformity and spacing.

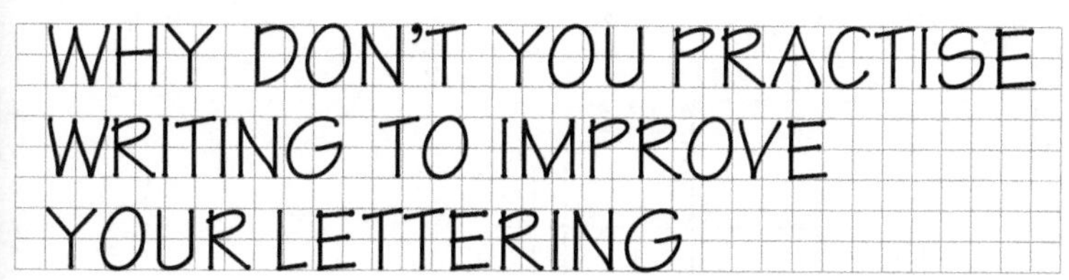

Figure 4.8

Practise writing your notes in Gothic vertical, and remember it is also important to follow the sequence of stroke indicated by the arrows in Figure 4.3.

Compare your lettering to that of a classmate and comment on the differences.

Keep some dated examples of your work in your portfolio of evidence.

BS EN ISO 10209 Technical product documentation – Vocabulary – Terms relating to technical drawings, product definition and related products

BS EN ISO 14660-1 Geometrical product specification (GPS) – Geometrical features – Part 1: General terms and definitions

5. Producing drawings for graphic communication

Option B Section 2.1

5.1 Dimensioning principles

International Standards Organisation (ISO) 128 specifies general rules and basic conventions which include standards and conventions for dimensioning drawings.

Clear and standardised dimensioning of technical drawings is essential for accuracy, clarity and unambiguity. This ensures that drawings are easy to read and understand, and that all the information needed is available (see Figures 5.1 and 5.2).

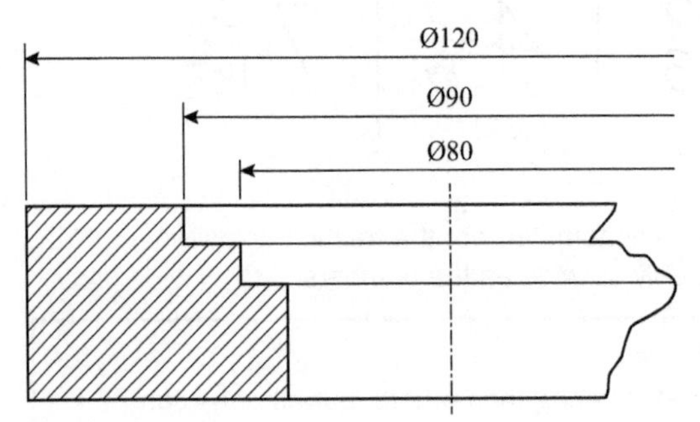

Figure 5.1 Technical drawing

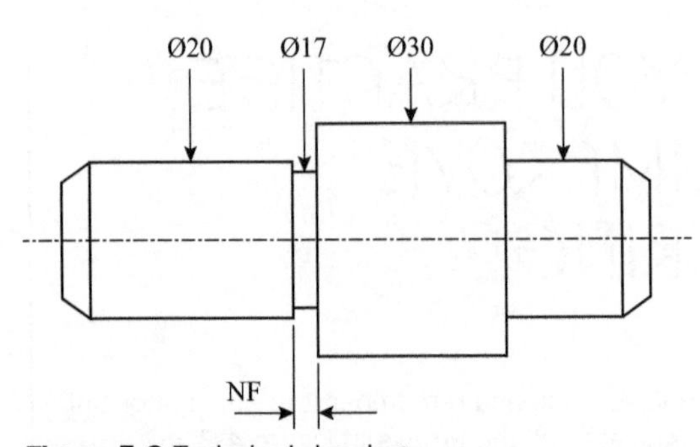

Figure 5.2 Technical drawing

Refer back to Section 3.2, page 26 for general principles of dimensioning.

Types of dimensioning

Parallel dimensioning

Parallel dimensioning is from a common feature where the dimensions are in the same direction originating from one projection line (see Figure 5.3).

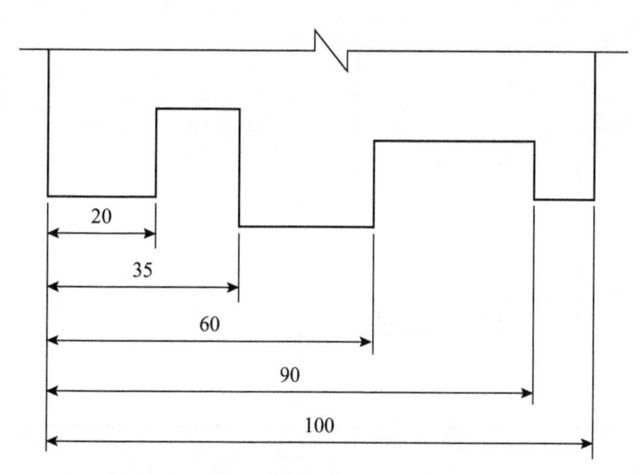

Figure 5.3 Parallel dimensioning

Superimposed dimensioning

Superimposed dimensioning is from a common origin marked (see Figure 5.4), and is a simplified version of parallel dimensioning. It is mostly used where there are space limitations on the drawing.

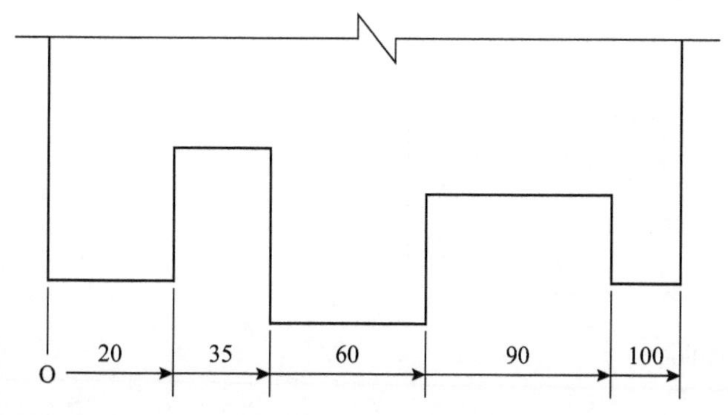

Figure 5.4 Superimposed dimensioning

Chain dimensioning

Chain dimensioning should only be used if the function of the component is not affected by any accumulation of the tolerances (see Figure 5.5). Tolerances are the allowable variations in the accuracy the component has to be made to.

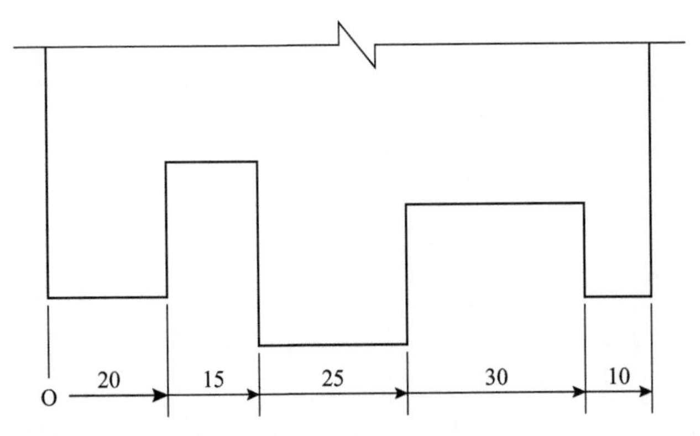

Figure 5.5 Chain dimensioning

See Section 5.3 Dimensioning conventions for tolerances, limits and fits.

Combined dimensioning

Combined dimensioning is a combination of parallel and chain dimensioning on the same drawing. Many organisations do not use combination dimensioning as it can cause confusion leading to mistakes. However, it can avoid accumulated tolerance error if used selectively.

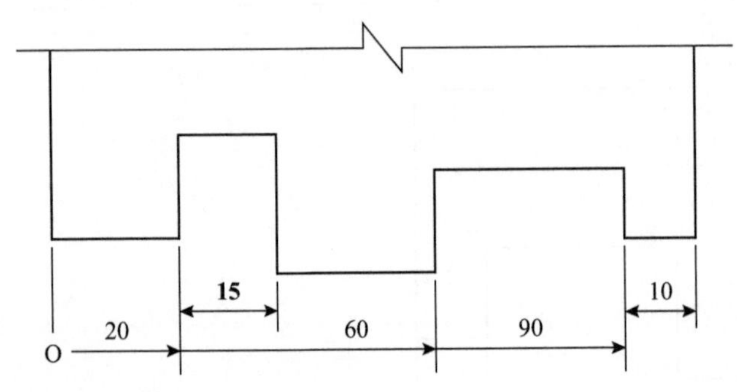

Figure 5.6 Combined dimensioning

5.2 Functional and non-functional dimensions

Remember dimensioning shows the size or value (in appropriate units) of a feature and can also indicate technical information through lines, symbols and notes.

Arrowheads are used to represent the two ends of the dimension line.

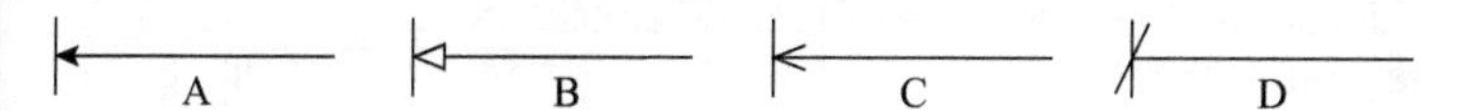

Figure 5.7 Dimension lines

Figure 5.7 shows the four most commonly used arrowheads in dimensioning:

- A is 30° closed and filled arrowhead (default).
- B is a 30° closed but unfilled arrowhead.
- C is 30° open arrowhead.
- D is simply an oblique stroke of about 30°.

Dimensions can be put into three main groups relating to the function of the component: functional dimensions, non-functional dimensions, and auxiliary dimensions.

Function dimensioning

Function dimensions (F) are essential as they directly impact on the functioning of the component (see Figures 5.8 and 5.9). They can affect the function in terms of size and/or location with its assembly within the finished product, and are usually toleranced. Consequently, functional dimensions are the subject of close manufacturing and inspection considerations.

Non-functional dimensioning

Non-functional dimensions (NF) are dimensions that are used for manufacturing purposes but do not directly affect the function or working of a component (see Figures 5.8 and 5.9). Non-functional dimensions are not usually subjected to very close inspection.

Auxiliary dimensioning

Auxiliary dimensions (AUX) are given for information only (see Figure 5.8). They are not used for manufacturing or inspection purposes, and are not to be toleranced. They are always inserted in parentheses (brackets).

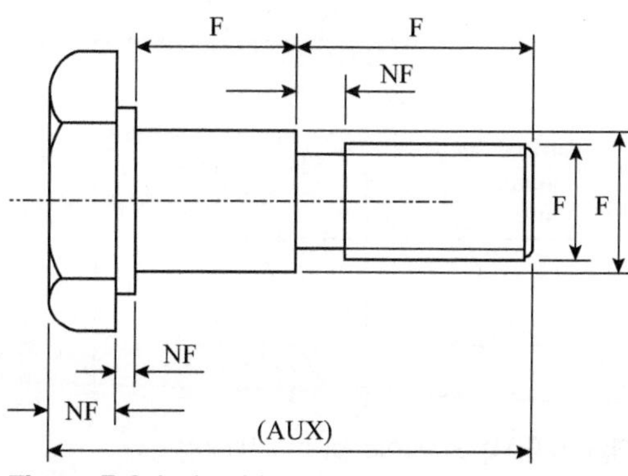

Figure 5.8 A shoulder screw

Assemblies

The flywheel (in sectional view in Figure 5.9) rotates on the shoulder screw. The internal diameter of the flywheel central boss and diameter of the shoulder screw are functional: the flywheel must rotate freely but without too much clearance, which would result in vibration and excessive wear. Likewise, the thickness of the central boss on the flywheel is functional as it must not bind on the engine body if it is to rotate freely. The drilled depth of the hole in the engine body to take the shoulder screw is non-functional. However the depth of internal thread in the same hole is functional as it will affect the amount of clearance on the flywheel central boss. To secure the required functionality of the features, tolerancing is needed.

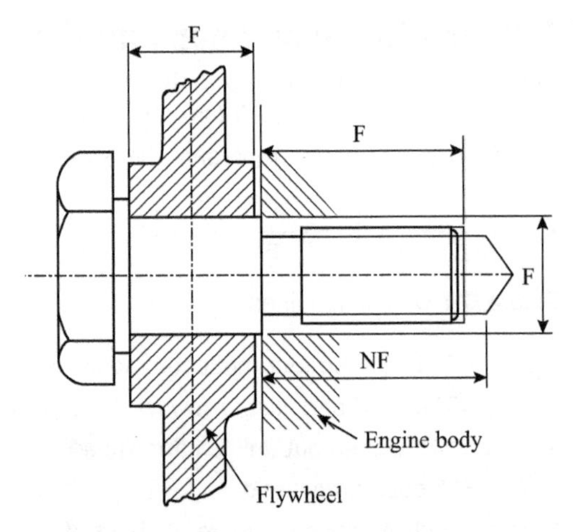

Figure 5.9 A shoulder screw in position

Surface finish

Manufactured components may have surface finishes from a range of processes: rolled, machined, stamped, painted, etc. The surface conditions can vary hugely. Surfaces may be functional or non-functional, but where a surface is functional this needs to be shown appropriately on the drawing. The requirements may be for machining for which the symbols are as shown in Figures 5.10, and Table 5.1.

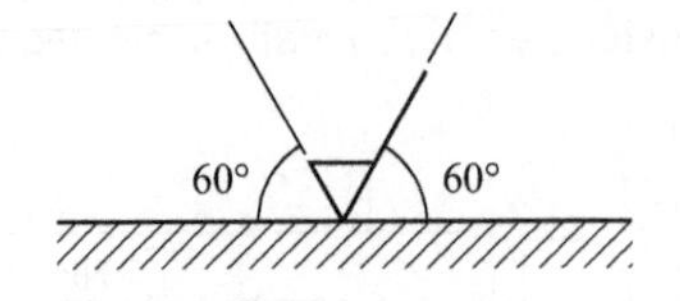

Surface can be produced by any process	Material removal required	Material removal prohibited

Figure 5.10 Symbols for machining

Table 5.1 Symbols and roughness values

When a particular surface finish is needed	
1. When the means of production are unimportant, this symbol is used	0.8
2. For a specific machining operation, this symbol is used	1.6
3. When no material can be removed, this symbol is used	3.2

Roughness value	General description
12.5	Rough machining
5.3	Course machining
3.2	Average machining
1.5	Good machining
0.8	Fine machining
0.4	Fine grinding
0.2	Honing
0.1	Buffing
0.05	Polishing
0.025	Super polishing

Dimensioning features

- Holes are dimensioned by giving their location and diameter in the circular view (see Figure 5.11).

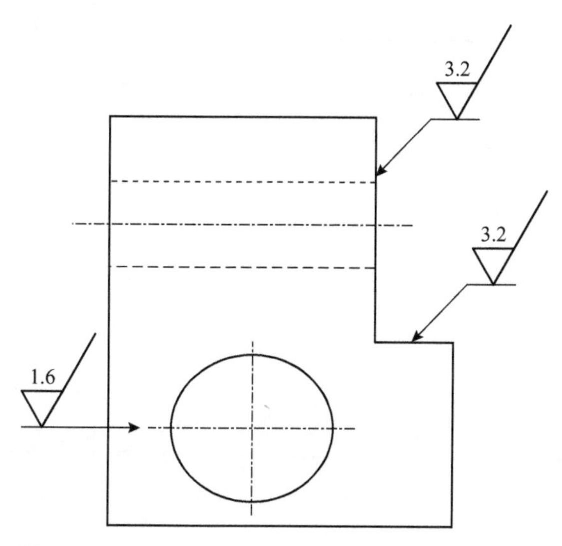

Figure 5.11 Dimensioning a hole

- Cylinders are dimensioned by their diameters and lengths in the side or rectangular view, and location in the circular view.
- For holes that go completely through the component and it is not clear on the drawing, the abbreviation "THRU" follows the dimension.
- The depth of blind holes can be specified in a note as dimensioning to hidden detail is poor practice.
- Symmetrical components (see Figure 5.12) can be drawn along the line of symmetry, with the symbol that looks like a curved equals sign turned vertically on the line of symmetry.

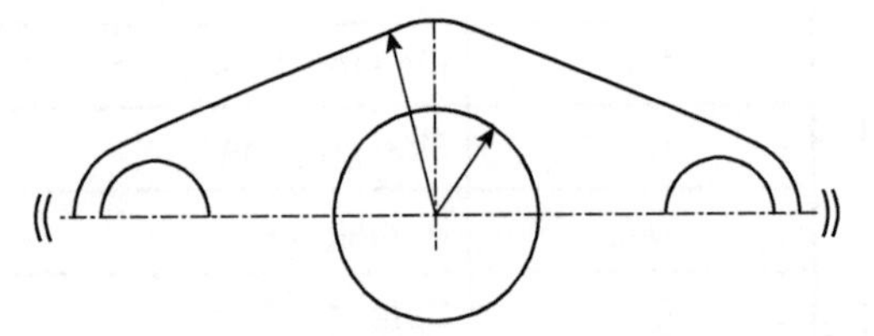

Figure 5.12 Symmetrical components

- Where a feature is repeated in a linear space (see Figure 5.13), it is permissible to use a simplified method of dimensioning. A note will help avoid any ambiguity.

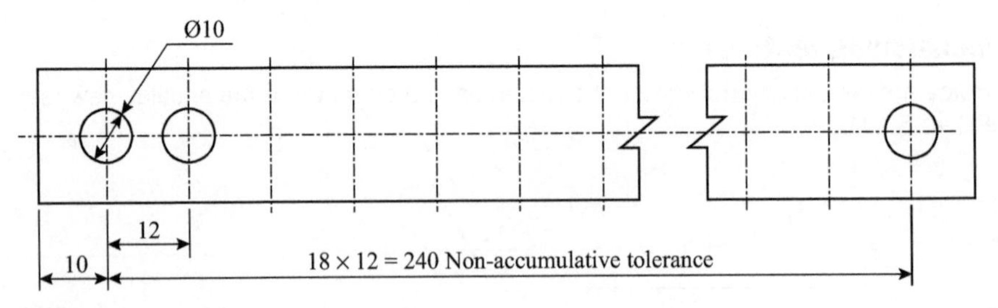

Figure 5.13 Repeating in a linear space

5.3 Dimensioning conventions for tolerances, limits and fits

Tolerancing is the means of communicating the upper and lower **limits** of the permissible variations in the final size, location and/or shape of a feature of a component. The difference between these upper and lower limits is called the **tolerance**.

The principle need for tolerances is that it is not possible to produce a component that is perfect, and therefore allowable variations are needed. The features of a component that are functional need to be made so that they function as required. This also means there is a high degree of uniformity of components required leading to interchangeability. Interchangeability of components is an important characteristic in the maintenance of mechanisms and systems.

Fit relates to the tightness or looseness of two mating parts, and how they move or rotate (function) relative to each other. Consider the difference between a wheel fixed to an axle and a shaft rotating in a bearing.

Note: These fits relate to features other than bearings and shafts. They equally relate to flat, square, and any other mating parts.

Limit dimensioning

There are three broad classes of fit:

- clearance fits
- interference fits
- transition fits.

Figure 5.14 shows a shaft and the bearing into which it fits. The smallest hole permissible diameter is 10.50, while the largest shaft diameter is 10.49. This means the shaft will always move or rotate freely in the bearing resulting in a **clearance fit** (smallest hole bigger than largest shaft).

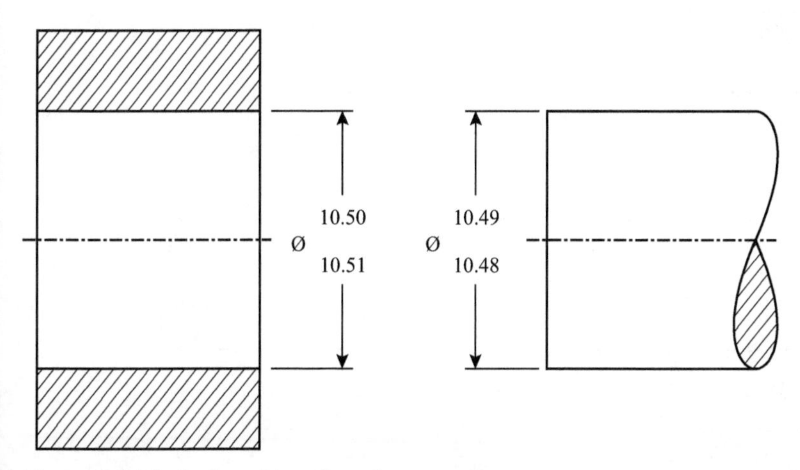

Figure 5.14 A shaft and bearing, clearance fit

Figure 5.15 again shows a shaft and the bearing into which it fits. However, the largest hole permissible is 10.49 and the smallest shaft 10.50. This means the shaft will not move or rotate freely in bearing resulting in an **interference fit** (largest hole smaller than smallest shaft).

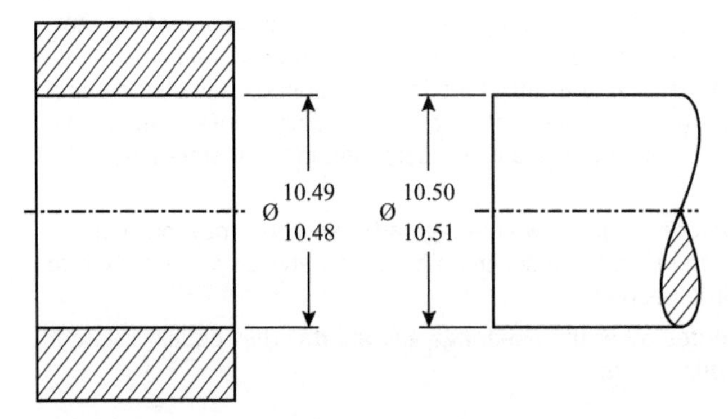

Figure 5.15 A shaft and bearing, interference fit

Figure 5.16 shows another shaft and bearing into which it fits. Two outcomes are possible here:

- If the largest hole (10.51) is fitted with the smallest shaft (10.49), a clearance fit is achieved.
- If the smallest hole (10.50) is fitted with the largest shaft (10.50), they are **"size and size"** meaning an interference fit is achieved. When both outcomes are possible this is a **transition fit**.

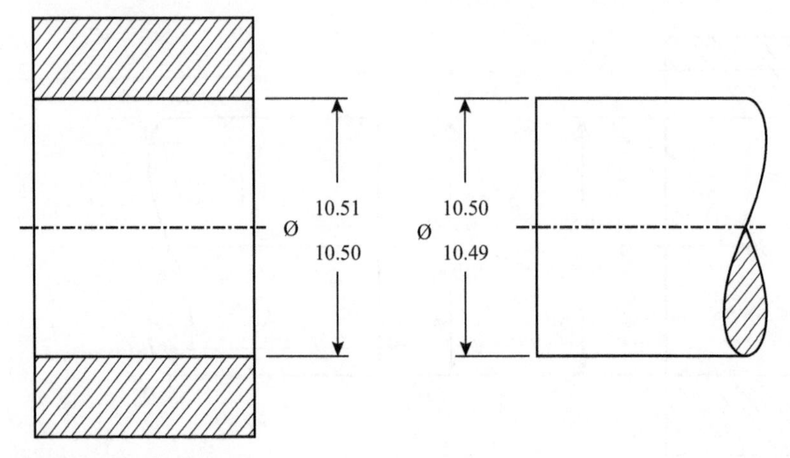

Figure 5.16 A shaft and bearing, transition fit

Dimensioning tolerances

Figure 5.17 shows a pulley with a keyway of nominal size 6 × 3 by which it will be keyed to a shaft. The width of the keyway is dimensioned with tolerance of 6.00–5.95. This is a **limit dimension** where the upper and lower limits are "stacked" on top of each other so anywhere between these two dimensions is permissible.

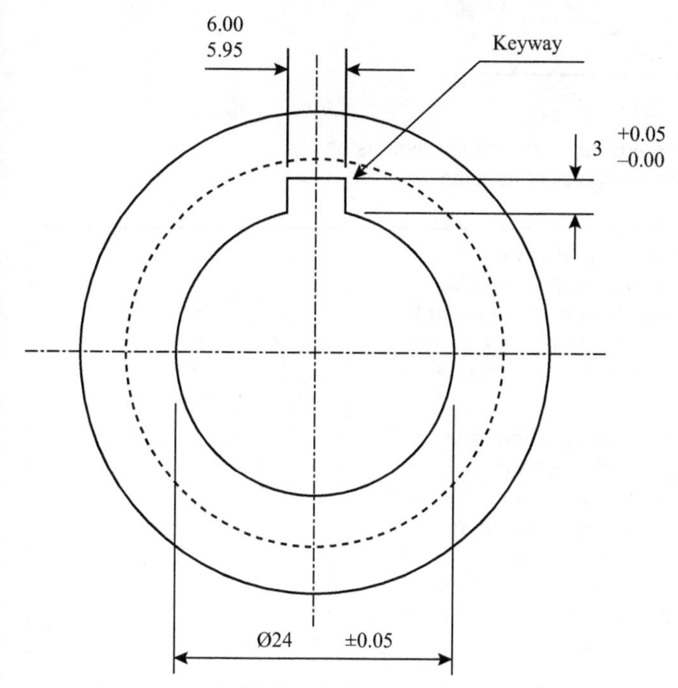

Figure 5.17 A keyway with a limit dimension

- The depth of the keyway is dimensioned 3 plus 0.05 and minus 0.00. This is a **unilateral tolerance** because the tolerance is on one side of the target dimension.
- The internal diameter of the pulley is dimensioned as Ø24 plus or minus 0.05. This is a **bilateral tolerance** as the limits are on both sides of the target of Ø24, so sizes Ø23.95 to Ø24.05 are permissible.

Remember:

- Tolerancing is the means of specifying the upper and lower limits of the permissible variation in the finished manufactured size of a feature. The difference between the upper and lower limits is the tolerance for that dimension.
- Functional and most non-functional dimensions (those that might affect interchangeability) are subject to tolerances but auxiliary dimensions are not.
- Dimensions, unless otherwise stated, are in millimetres (symbol mm) so there is not the need to include this on drawings.
- Tolerances, unless otherwise stated, are linear ±0,4 and angular ±0° 30'.

Tolerancing angular dimensions

Bilateral tolerancing of an angular dimension is shown in Figure 5.18 (a).

Unilateral tolerancing of an angular dimension is shown in Figure 5.18 (b).

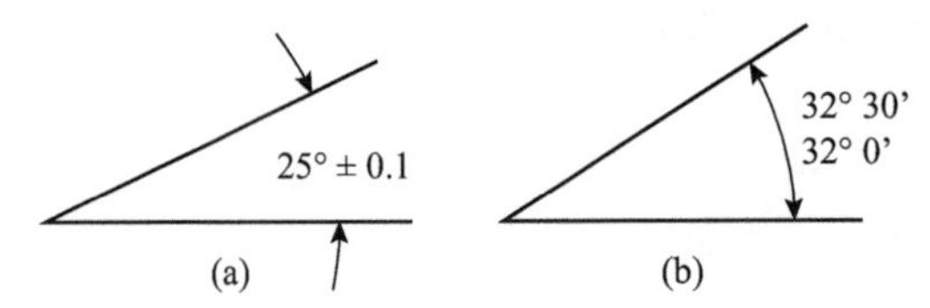

Figure 5.18 (a) bilateral tolerancing of an angular dimension (b) unilateral tolerancing of an angular dimension

Why don't you?

1. Draw the un-dimensioned drawing (Figure 5.19) of a bearing housing that is accurately located on a "dovetail" slide, and located by two holes to the machine bed. The bearing housing carries a shaft that rotates at high speed.

Draw the component and dimension those features that you think are functional F, non-functional NF, and auxiliary (AUX). Also show the surface finish which may be appropriate for each functional dimension. Once you have dimensioned the drawing, state why you have decided on each dimension and surface finish.

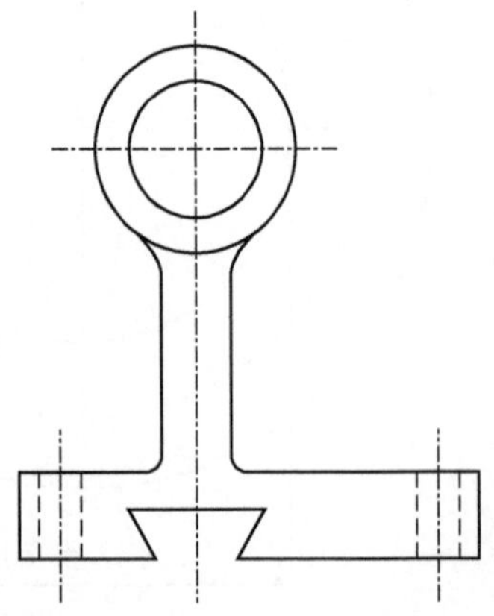

Figure 5.19 A bearing housing

2. Figure 5.20 shows two mating parts: a shaft and a hole. Study the figure, and then answer these questions.

a) What class of fit is it?

b) What are the tolerances on the hole and the shaft?

c) Are the tolerances unilateral or bilateral?

d) What is the maximum and least material condition (MMC and LMC) for both components?

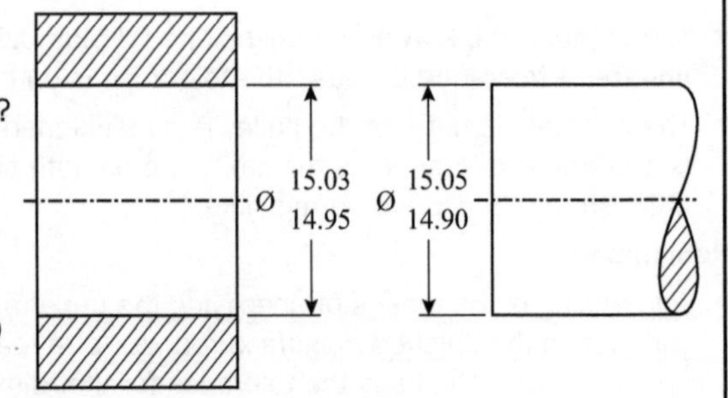

Figure 5.20 Two mating parts

Check your work with a classmate. Discuss what went well and what could be improved.

Put a signed and dated copy of your work in your portfolio of evidence.

6. Producing pictorial and orthographic drawings

Option B Section 2.2 & 2.3

6.1 Characteristics and use of isometric and oblique drawings

Pictorial drawings are projections of a two-dimensional illustration of a three-dimensional object. Because they show three faces of the object, they give a realist view and can be some of the easiest drawings to understand.

There are two main types of pictorial drawings: **oblique** and **isometric** (oblique pictorials can be either cavalier or cabinet).

Figure 6.1 shows three drawings of the same component using different pictorial drawings.

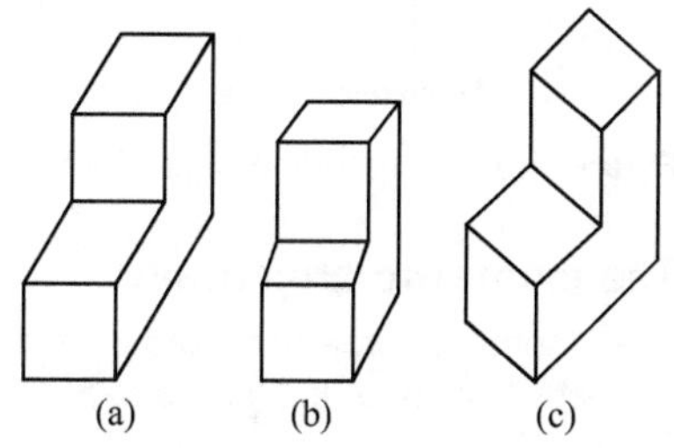

Figure 6.1 (a) Cavalier, (b) cabinet, and (c) isometric projections

Oblique projections show the front surface in its true square shape, while the other surface (top and side) slopes at an angle of 45° or 30°. The difference between **cavalier** (see Figure 6.1(a)) and **cabinet** (see Figure 6.1(b)) projections is that cavalier oblique sides are drawn full length, while cabinet oblique sides are reduced to half length.

Note: The reduced side is still dimensioned to full size. Cabinet oblique projections are sometimes preferred because they give a more realistic view.

Isometric (meaning equal measure) projections show both sides sloping at 30° or 45° while the vertical remain vertical, as in Figure 6.1(c).

It is sometimes helpful to think about pictorial drawings using oblique and isometric axes. Figures 6.2 and 6.3 show the receding angles for both oblique and isometric projections. Sometimes the receding angle is 60°.

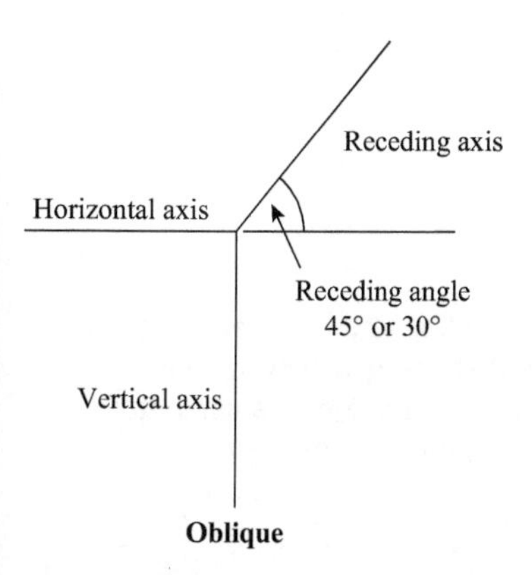

Figure 6.2 The receding angle for oblique projections

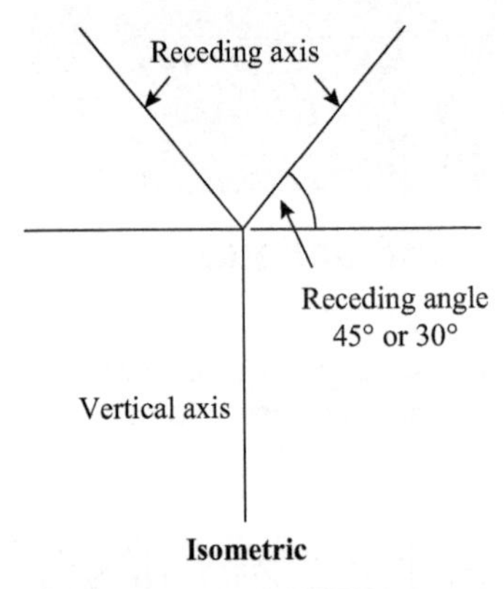

Figure 6.3 The receding angle for isometric projections

The isometric projection

In an isometric projection, the axes are more correctly known as the *x*-, *y*- and *z*- axes. In Figure 6.4:

- the *x*-axis is 30° to the horizontal
- the *y*-axis is 30° to the horizontal
- the *z*-axis is 90° to the horizontal, and is therefore vertical.

These axes will become more relevant later.

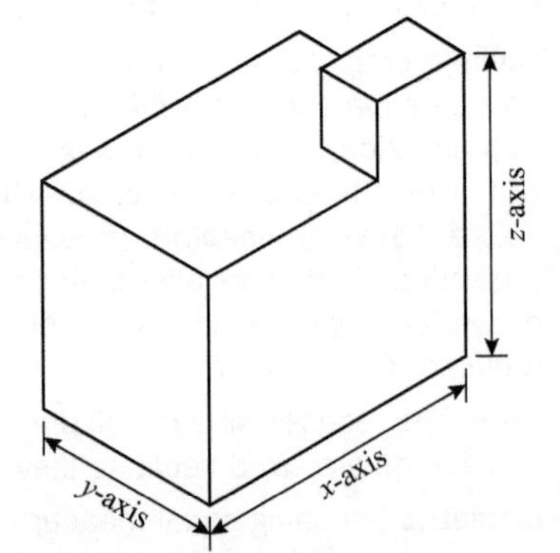

Figure 6.4 The axes in an isometric projection

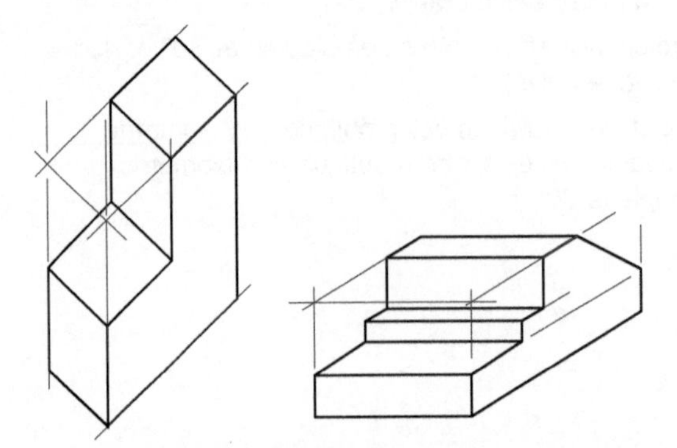

Figure 6.5 Examples of isometric projections

The approach to making a pictorial drawing is to produce a box to the overall dimensions of the object and then remove portions until the required shape is achieved. This produces clearer images and is more time efficient. If you are producing a cabinet oblique, remember to draw the box half ($\frac{1}{2}$) depth of the actual size (or scale).

6.2 Producing isometric drawings of regular and irregular objects

Positioning

Isometric axes can be positioned to give different views of the same object. Figure 6.6(a) shows a regular isometric, in which the viewpoint is looking down on the top of the object. The axes at 30° to the horizontal are drawn upward from the horizontal. Figure 6.6(b) shows a reversed axis isometric, the viewpoint is looking up on the bottom of the object, and the 30° axes are drawn downward from the horizontal. Figures 6.6(c) and (d) show the long axis isometric, where the viewpoint is looking from the right and left of the object, and axes are drawn at 60° to the horizontal.

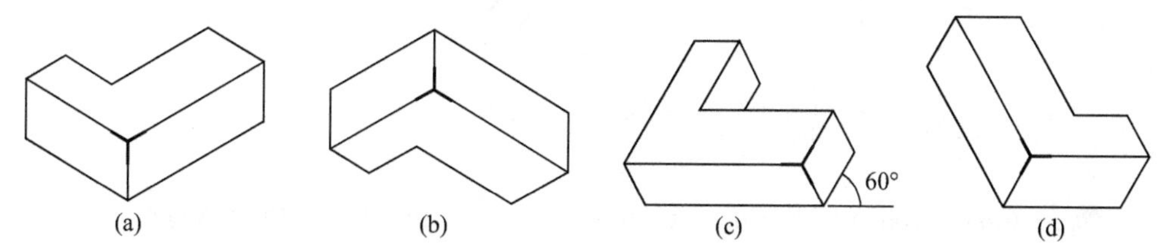

Figure 6.6 Positioning isometric axes

Isometric drawing

The box method for creating isometric drawings has four stages:

1. Identify the isometric viewpoint that best shows the features of the object, then draw the isometric axes which will create that view.
2. Draw the isometric planes, using the overall width, height, and depth of the object, so that the object is enclosed in a box.
3. Locate details on the isometric planes.
4. Darken all visible lines – hidden detail is not shown unless absolutely necessary to fully describe the object.

Isometric and non-isometric lines and planes

- **Isometric lines** are lines that run parallel to any of the isometric axes (see Figure 6.7).
- **Non-isometric lines** are any lines that are not parallel with the isometric axes.
- Non-isometric lines include inclined lines, and they must be produced by locating two end points.
- Surfaces formed by any two adjacent isometric axes are **isometric planes**.
- Planes that are not parallel to any isometric plane are called **non-isometric planes**.

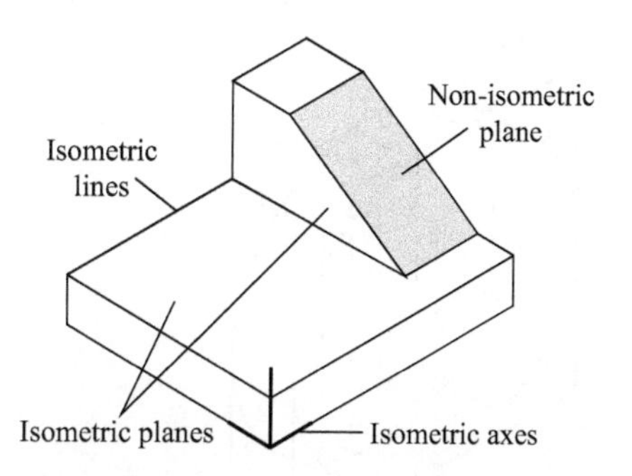

Figure 6.7 Isometric and non-isometric lines and planes

Objects with inclined surfaces

In Figure 6.8(a) an octagon has been drawn in isometric projection. The non-isometric lines and planes have been constructed by finding the two end points for the inclined surface by measuring *a* and *b* from the front elevation (Figure 6.8(b)) of the octagon and then transferring them to the isometric projection. Because the figure is an octagon, lengths *a* and *b* are equal.

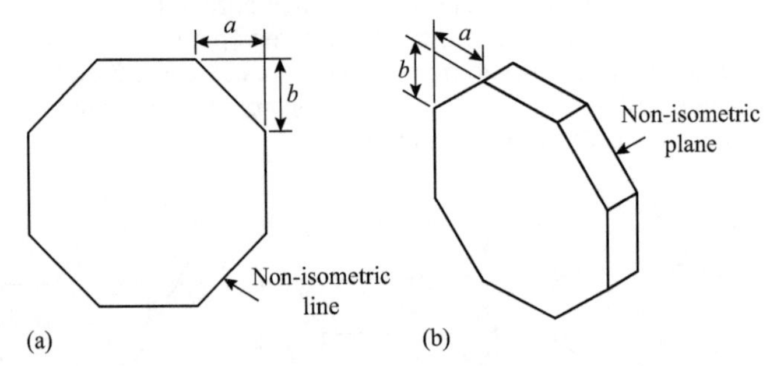

Figure 6.8 (a) Isometric drawing of an octagon (b) front elevation of the octagon

Drawing circles and cylinders

Figure 6.9 shows a circle drawn in isometric projection.

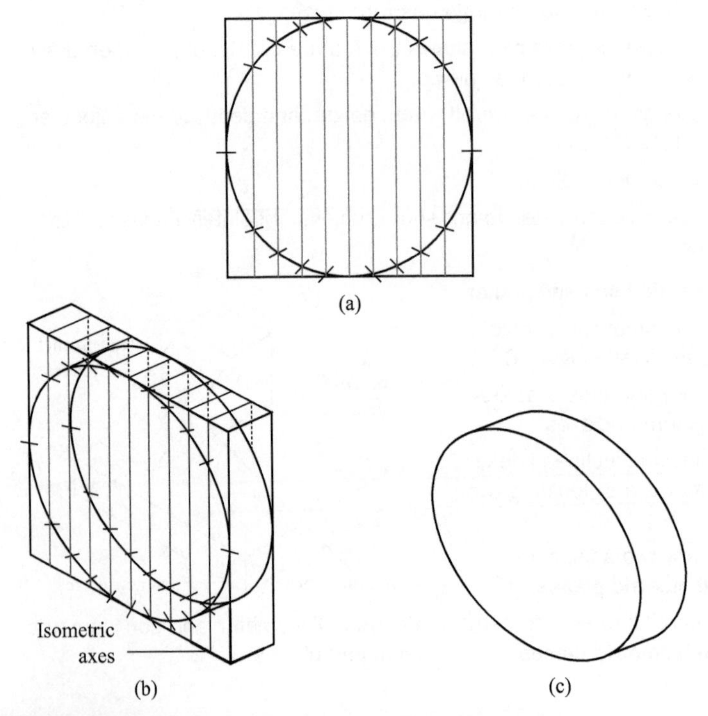

Figure 6.9 Stages in drawing an isometric circle

The circle is first drawn in front elevation within a square and divided into equal parts vertically as in Figure 6.9(a). The outline box is drawn isometrically and divided vertically (both front and back surfaces) into the same number of equal parts. From the front elevation, the intersection lines of the circle's circumference are measured from the top and bottom of the square, and then transferred to the corresponding division line on the isometric surfaces. The intersections are then joined up by a smooth arc to create the isometric circle which appears as an ellipse as in Figure 6.9(b). The two circles are joined together to form a cylinder and the construction lines are removed, as in Figure 6.9(c).

Irregular curves in isometric drawing

The irregular shape is drawn in front elevation and divided into a number of equal parts vertically as in Figure 6.10(a). The outline box is drawn isometrically and divided vertically into the same number of equal parts. The intersections of the component profile with the vertical lines on the front elevation are measured and transferred to the corresponding lines on the isometric projection, and then joined together with a smooth curve as in Figure 6.10(b).

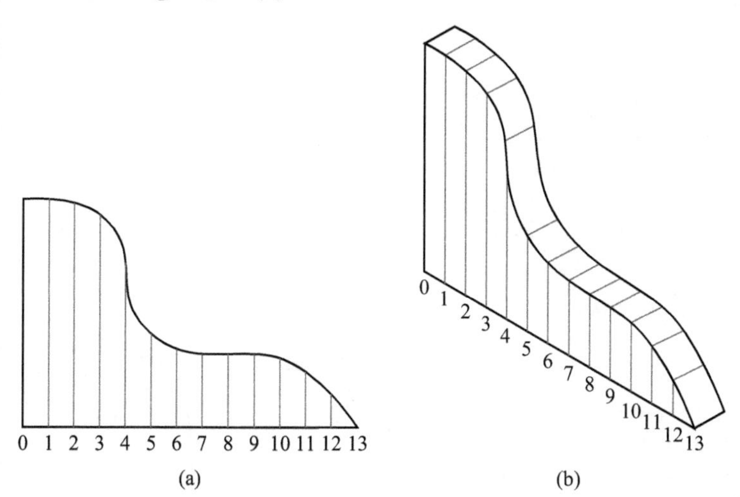

Figure 6.10 Stages in drawing an irregular curve

6.3 Oblique drawings with curves

Key features of oblique pictorial views

Producing oblique drawings:

1. Oblique is most suitable for simple components.
2. Visualise the component. The front surface and surfaces parallel to it are shown in true right angles, i.e. viewed straight on.
3. Choose the most appropriate surface to have parallel to the plane of the project (the one facing outwards and square on). This normally includes the longest surface and/or surfaces with circles and/or arcs.

4. Minimise the amount of distortion of the image by choosing the most appropriate receding axis angle (usually those with lower receding axis offer less distortion).
5. Decide whether oblique cabinet or cavalier is best.
6. Centre lines are not used unless necessary for dimensioning.
7. Likewise hidden detail is not used unless necessary to fully describe the component.
8. Draw a box that will just fit the component then build up the detail within the box.

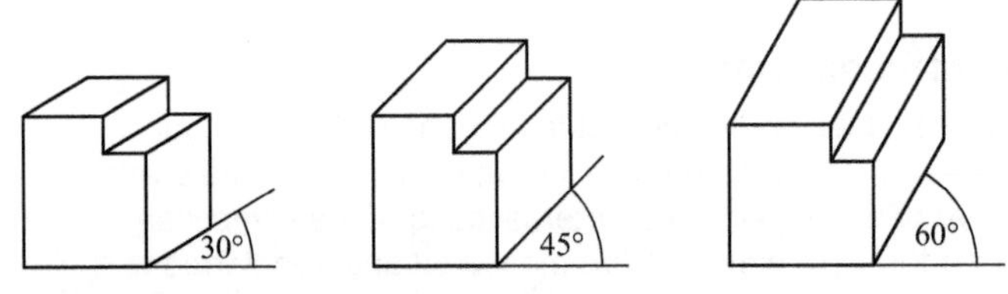

Figure 6.11 Variations in oblique angles

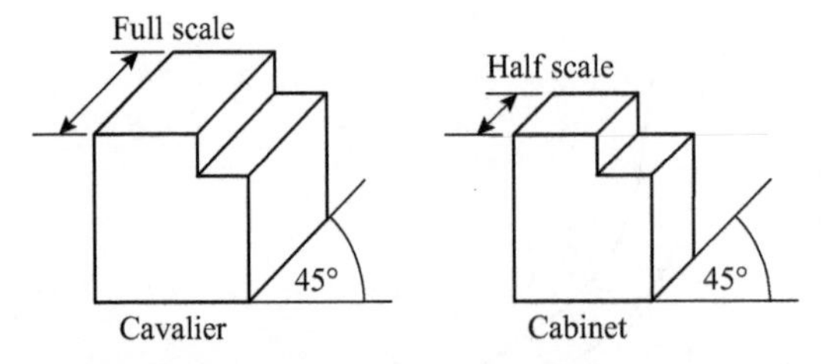

Figure 6.12 Types of oblique drawing

Drawing circles

In Figure 6.13 the component has been drawn in a box and then the detail built up. Consider how difficult and time consuming it would be if the circles were drawn on the receding axis rather than the horizontal one.

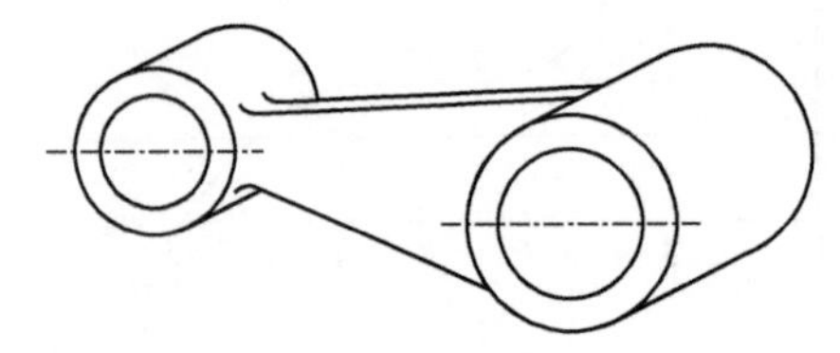

Figure 6.13 A component

Note: The centre lines have been included to show that the two holes are not on the same centre line to best describe the component. Centre lines and hidden detail are not usually shown on oblique views.

Components with more than one arc

Sometimes it is necessary to draw circles and arcs on the receding axis.

In Figure 6.14 a component with two arcs (semi-circles) has been drawn in oblique cabinet. As there is an arc at each end of the component, one has had to be drawn on the receding axis. Because the receding axis is in cabinet (half-size), the arc on that axis has had to be fore-shortened by half.

Note: The arc is only fore-shortened on the receding axis, while it is full size on the vertical axis.

The arc on the horizontal axis has been divided into 12 equal parts. The length of each division is measured and then halved, that length is transferred to corresponding lines on the receding axis arc then joined together with a smooth curve.

Oblique cabinet

Figure 6.14 A component

Components with circles

In Figure 6.15 a block has been drawn in oblique cabinet with a circle on both the horizontal and receding axes.

The circle on the front face has been divided into nine equal parts vertically. These nine divisions have been transferred to the circle on the receding axis but because the axis is half full size, the distance between each vertical line has been halved. The length of the vertical lines is full size and then the intersections are marked off and transferred to the corresponding line on the receding axis circle as shown. The fore-shortening of the distance between the vertical lines and the full size of the vertical measurements gives the circle the appearance of an ellipse.

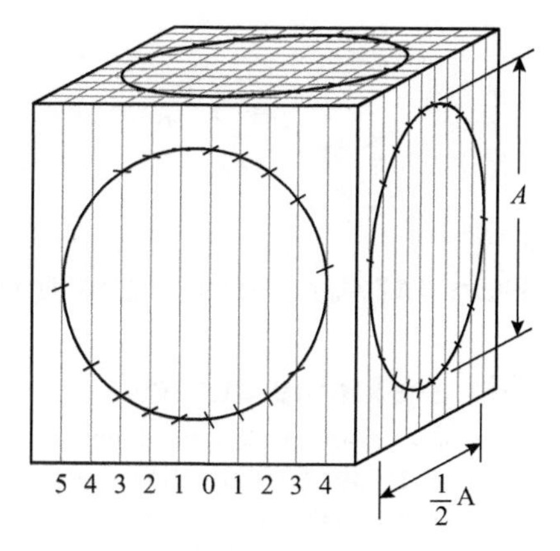

Figure 6.15 A circle on both the horizontal and receding axes

Combination of features

In Figure 6.16 the component is drawn in oblique cavalier so the 30° receding axis is full size. The component has two rounded corners of the same radius and has been drawn with axes that best describe the component.

A circle of the same radius was drawn and a quadrant of it divided into equal parts as was each of the rounded corners. The length of each line was transferred from the quadrant to the corresponding line on the rounded corner and then joined with a smooth curve.

The angular surfaces were marked out by finding two points then joining them with a straight line.

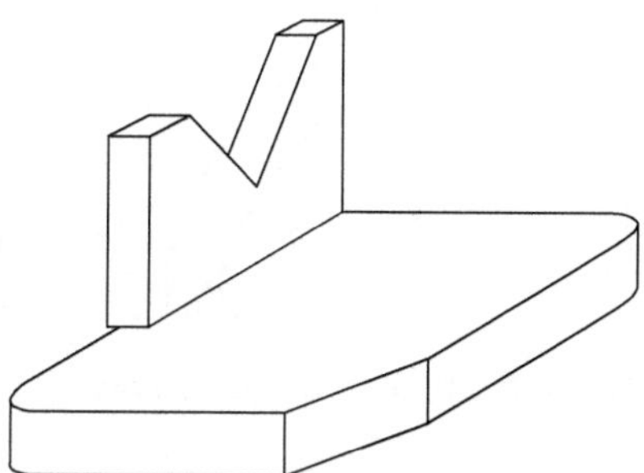

Figure 6.16 A component

6.4 Dimensioning pictorial drawings

Isometric dimensioning

The dimension lines are drawn parallel to the isometric axes and the extension lines are extended out from these axes. Dimensions can be:

- **aligned** – where the dimensions are aligned with the dimension lines, as in Figure 6.17(a)
- **unaligned** – (unidirectional) where the dimensions are all printed the same way, usually horizontally as in Figure 6.17(b).

It is not good practice to mix aligned and unaligned dimensions.

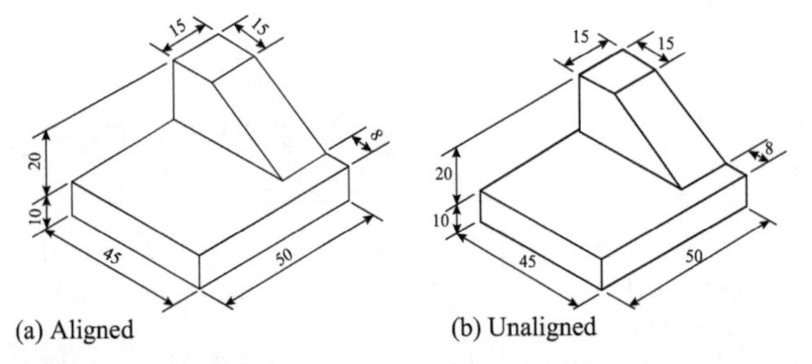

Figure 6.17 Aligned and unaligned dimensions

Dimensioning curves

Curves, arcs and circles are dimensioned in pictorial views as they are in other drawings. Figure 6.18 demonstrates how much more convenient it is to use oblique projection when drawing curves, arcs and circles, because they are drawn on the front face. There is less distortion, and the time required to produce the drawing is reduced.

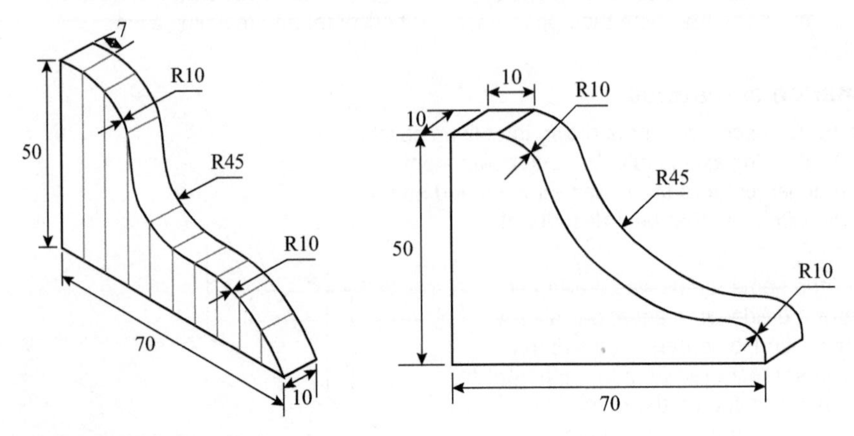

Figure 6.18 Dimensioning curves

Oblique dimensioning

In oblique dimensioning, all dimension lines and extension lines must be in the same plane as the feature to which they relate. Both aligned and unaligned are acceptable, as in Figure 6.19.

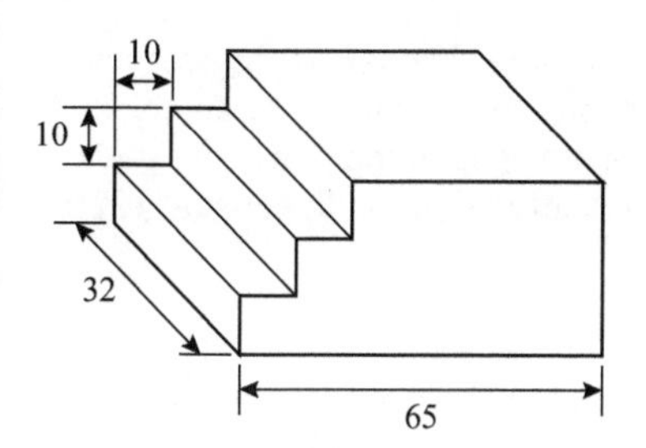

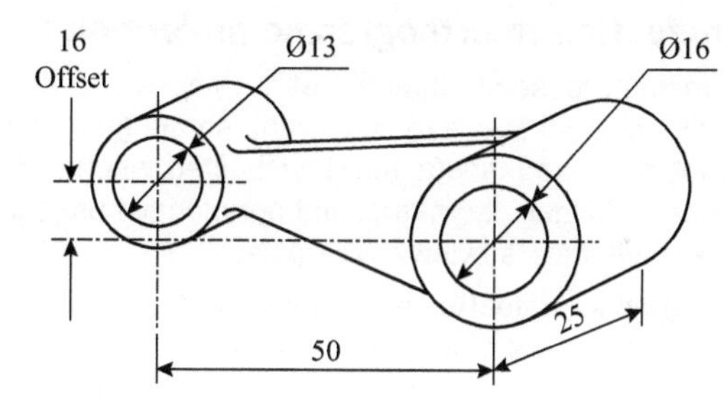

Figure 6.19 Oblique dimensioning

Why don't you?

1. The oblique cavalier projection in Figure 6.20 is drawn full size.
   - **a)** Redraw the component in oblique cavalier three-quarters ($\frac{3}{4}$) full size.
   - **b)** Redraw it in oblique cabinet.
   - **c)** Also redraw an isometric projection with the receding axis 30°.
   - **d)** On the isometric drawing label the *x*-, *y*- and *z*-axes.
   - **e)** Dimension all necessary features.

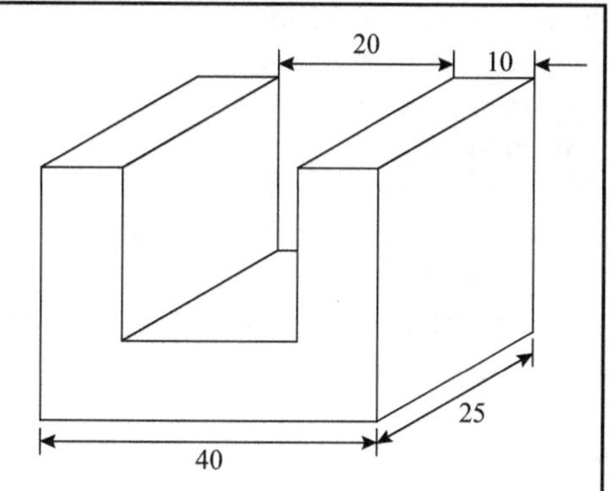

Figure 6.20

2. The component shown in Figure 6.21 is in long axis isometric view.
   - **a)** Redraw it in regular isometric projection and show all your construction lines (outline box). You will need to draw it in front elevation first.
   - **b)** Indicate the isometric axes used.
   - **c)** Identify any non-isometric lines and/or planes.
   - **d)** Should any centre lines be shown on the drawing?

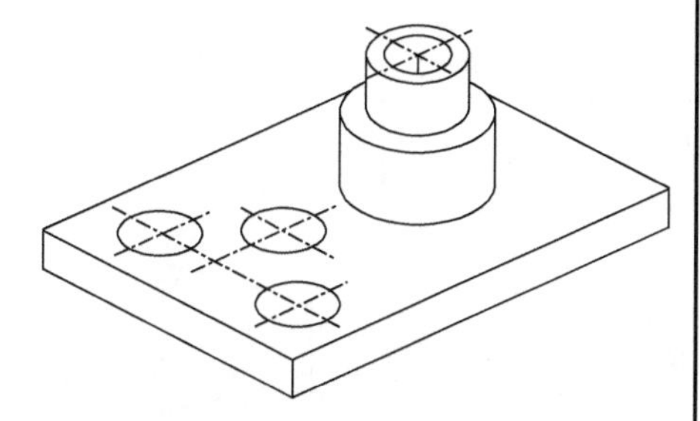

Figure 6.21

7. First and third angle projections

Option B Section 2.2 & 2.3

7.1 Principles of first and third angle planes of projections

Introduction to orthographic projection

The main purpose of all technical drawing is to represent three-dimensional objects in a two-dimensional space to convey the shape (geometry), size, and structure of an object so that it can be manufactured, inspected, and later maintained, as the designers had intended. Oblique, isometric, and perspective projections attempt to do this, usually in a single, multi-faceted view of the object.

Orthographic projection uses a series of two-dimensional views arranged in a standard layout to fully describe the geometry, size, and structure of the object. It does this by the creation of two-dimensional views of the object from three mutually perpendicular planes (see Figure 7.1). Each of the views is drawn to the same scale.

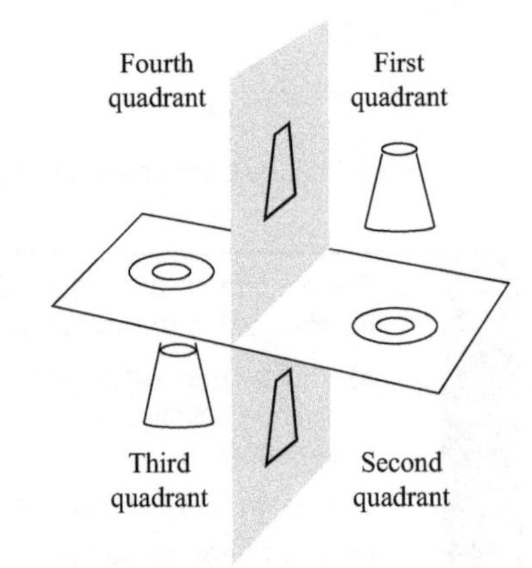

Figure 7.1 The four quadrants

- Projections that use the views in the first quadrant are referred to as **first angle projections**.
- Projections that use the views in the third quadrant are referred to as **third angle projections**.
- First angle projection had its origin in Europe.
- Third angle projection had its origin in North America.
- Both systems have international approval and are considered to be of equal status.
- There is an agreed symbol to convey which angle of projection has been used on a specific drawing, as shown in Figure 7.2.

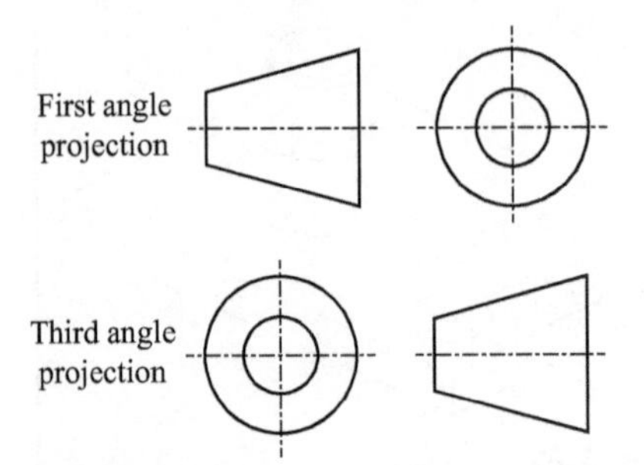

Figure 7.2 Symbols to show angles of projection

First angle projection

To produce a first angle projection drawing:

Step 1: Draw an elevation – usually the front elevation.

Step 2: Look at the front elevation in the direction of the arrow A, see Figure 7.3, and draw the side elevation.

Step 3: Look at the top of the front elevation in the direction of arrow B, and draw the plan view.

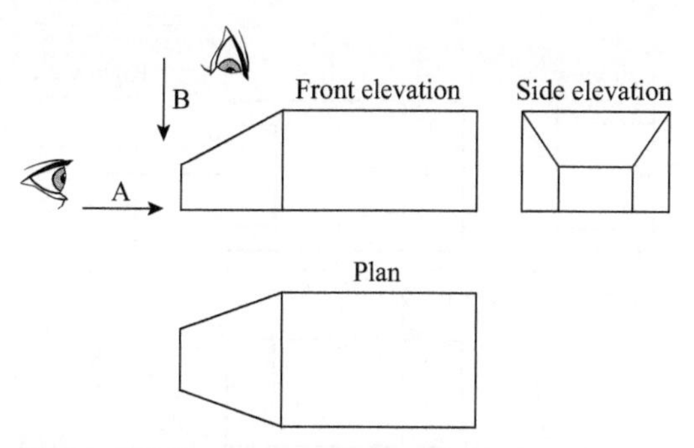

Figure 7.3 Producing a first angle projection drawing

Third angle projection

To produce a third angle projection:

Step 1: Draw an elevation – again the front elevation.

Step 2: Looking at the front elevation in the direction of the arrow A, see Figure 7.4, draw the side elevation to the left.

Step 3: Looking at the front elevation in the direction of arrow B, draw the plan view above.

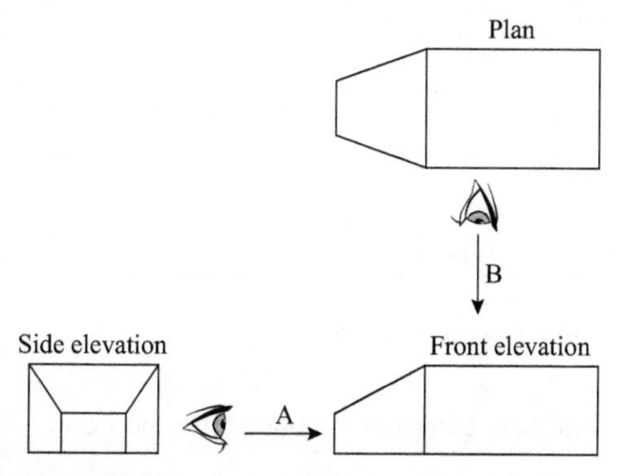

Figure 7.4 Producing a third angle projection drawing

Whether first or third angle projection is used, the guidance is:

- Choose a view that best describes the object.
- Project other views from this view.
- Profile lines of the object have precedence over all other lines.
- Hidden detail and section lines have precedence over centre lines.
- Centre lines have lowest precedence.

Transferring features

In Figure 7.5, a component has been drawn with the front view first, the view that best describes the component. The other views are then projected from that view with detail added in the form of additional views: right- and left-hand views. These are transferred using T and set squares including hidden detail and centre lines. The profile of the plan view can be transferred using a 45° set square and vertical lines, as shown, with hidden detail coming from the front view. Alternatively compasses can be used instead of a 45° set square.

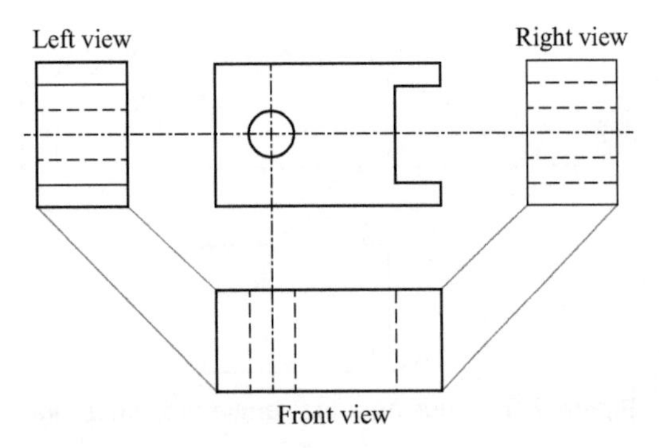

Figure 7.5 A component

Note: The maximum number of views used is normally three. More are used only if clarity is not given by three views. Sometimes a pictorial view is also included.

In Figure 7.6 a component has been drawn with the front view that best describes the component, then the other views have been produced by projecting features from this view including hidden detail and centre lines. The transfer of detail to the plan view from the front view has been done using compasses.

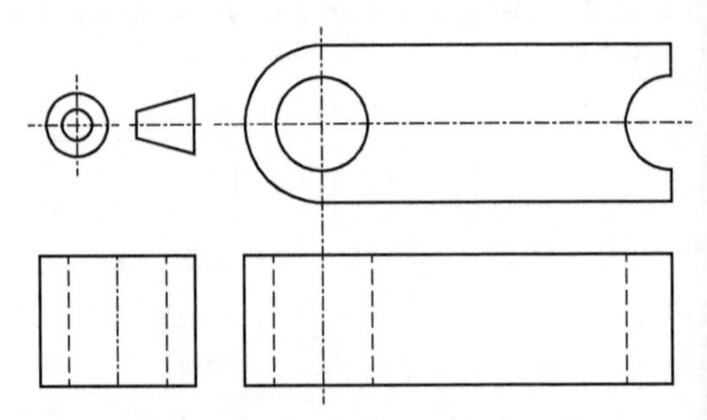
Figure 7.6 A component

It is **important** to remember:

- Never mix first and third projections.
- The symbols for first and third projections are only shown in the title block and never on the drawing itself.
- The scale used on orthographic drawings is also only shown in the title block.

7.2 First and third angle projections of machine assemblies and components

Vee blocks

Vee blocks are widely used in workshops for marking out and for inspection purposes, including checking for roundness of cylindrical workpieces. Generally the angle of the "V" is 90°. They can be clamped to a machine table for holding cylindrical workpieces.

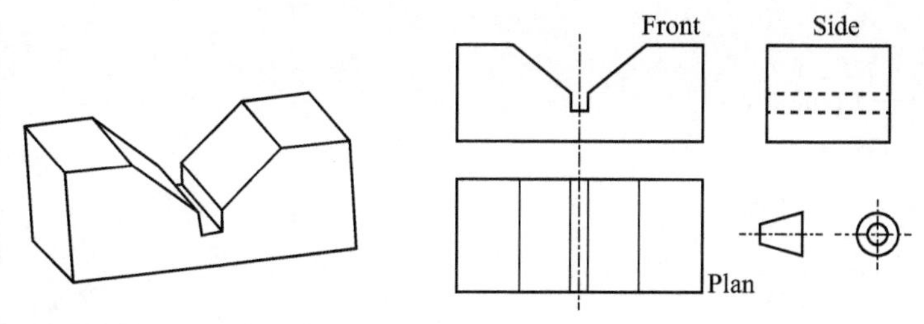

Figure 7.7 Vee blocks

Universal coupling

Couplings are components that connect two shafts. Their main role is the transmission of power to rotating parts. Universal couplings (such as in Figure 7.8) connect two shafts that may not be in perfect axial alignment but still transfer the torque and direction of rotation.

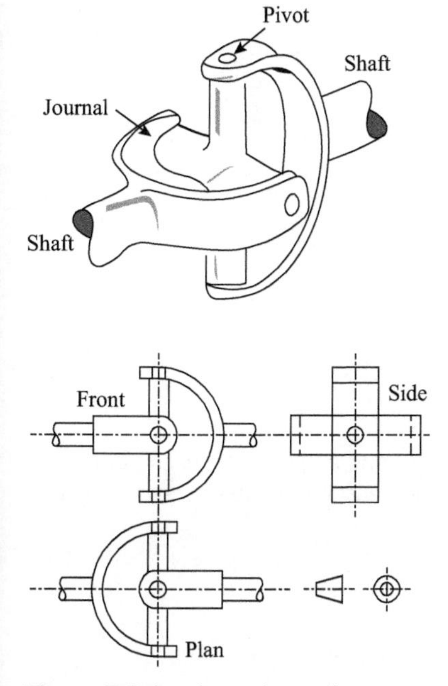

Figure 7.8 A universal coupling

Toolmaker's clamp

Most toolmaker's clamps are made of just four components as shown in Figure 7.9:

- an upper jaw
- a lower jaw
- a centre screw
- an outer screw.

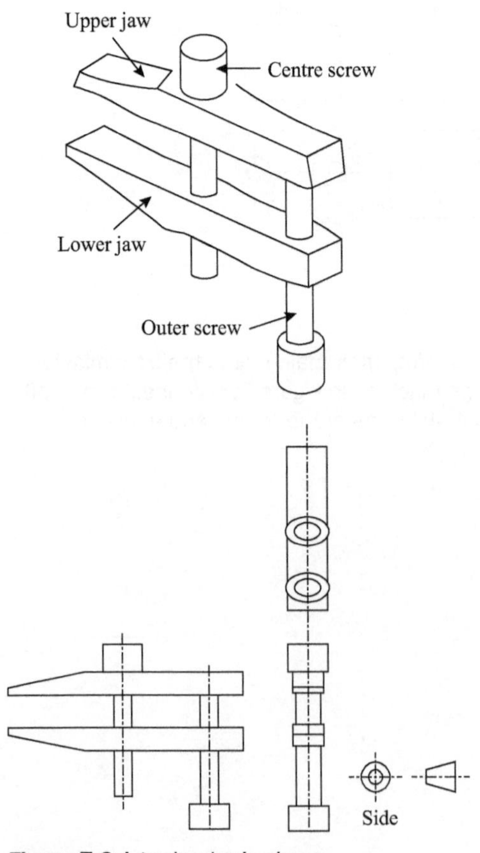

Figure 7.9 A toolmaker's clamp

The upper jaw has a clearance hole for the centre screw and a blind hole for the outer screw. The lower jaw has two screwed holes to take the centre and outer screws.

Why don't you?

1. Using the pictorial drawing (Figure 7.10):

   a) Redraw the component in both first and third angle projections.

   b) Produce and complete a title block for each of the drawings, correctly positioning the symbols for first and third angle projections and the scales used.

   Note: It is not necessary for you to dimension your drawings at this time.

   c) State the projection of the original pictorial drawing.

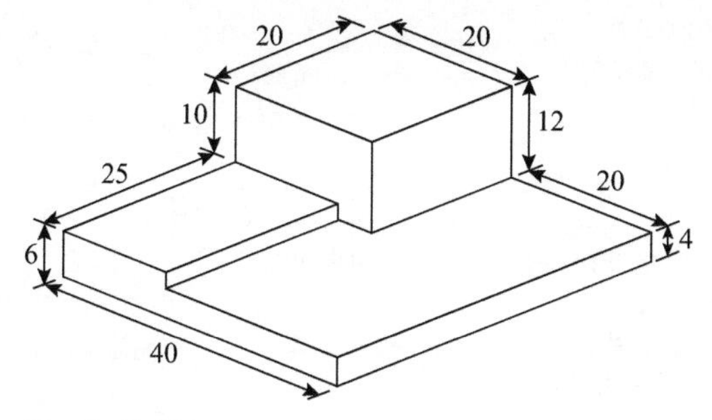

Figure 7.10

2. Draw the engineer's square shown in Figure 7.11 to 1:2 scale in first or third angle projection. Include a title block and show the angle of projection used.

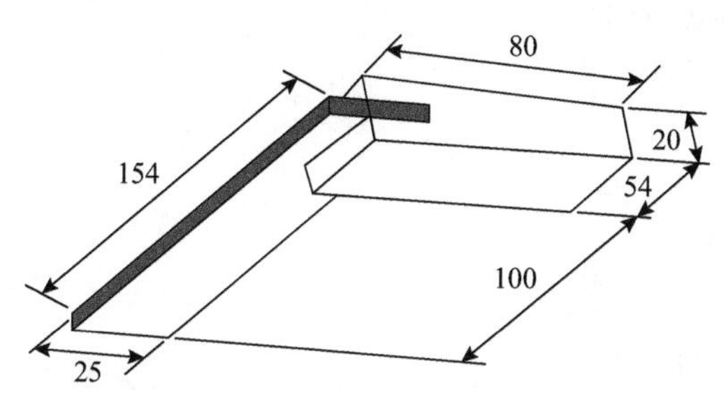

Figure 7.11

Put a copy of your work in your portfolio of evidence.

8. Centre lathe

Option B Section 3.6.6d

8.1 Setting up for a range of turning operations

Setting tool set at centre height

Figure 8.1 (a) shows a piece of round bar being machined with the cutting tool on centre height. The point of the tool is on the centre line passing through the workpiece. The cutting tool angles play a vital role in surface finish and the efficiency of the cutting action. This ensures the top rake (angle B) on the tool is ground to suit the material being machined.The clearance angle A is formed from the underside of the tool and a tangent of 90^0 to the centre line at the point of contact between the tool and the circumference of the workpiece.

Figure 8.1 (b) shows the cutting tool set above centre height. The rake angle B is increased, while the clearance angle A is significantly reduced. This setting will not only impact negatively on the surface finish but also heat generated (rubbing) and reduction on tool life.

Figure 8.1 (c) shows the cutting tool below centre height. The rake angle is reduced and the clearance angle increased. Again the surface finish is rougher, more heat is generated (greater cutting forces), and tool life reduced.

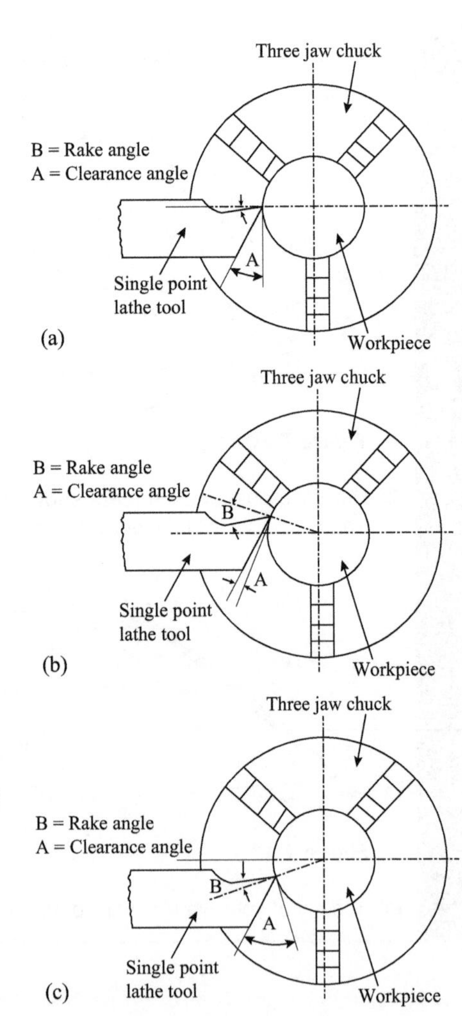

Figure 8.1 Changing angles of the lathe

See more on the rake angle in Section 2.1, page 14.

Setting the cutting tool at centre height

- If a tailstock centre is placed in the tailstock it is a convenient reference point to set the cutting tool (see Figure 8.2).
- The compound slide needs turning so the cutting point of the tool is presented to the tailstock centre.
- Packing may be needed (if the lathe does not have an adjusted tool post) to set the point of the cutting toll in line with the centre in the tailstock.

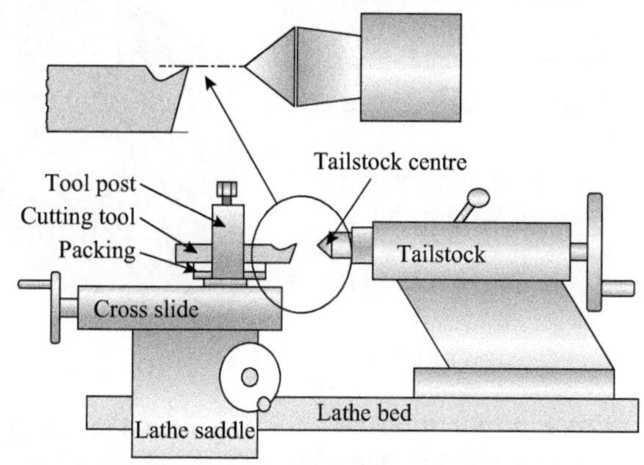

Figure 8.2 Setting the cutting tool at centre height

- Once set at centre height, the rake and clearance angles will be set as ground.
- If the cutting tool needs to be reground during machining, it will be necessary to check the centre height of the tool and adjusted if needed.

Turning a diameter between centres using catch plates

- Remove any work hold device, such as a three- or four- jaw chuck.
- Position catch plate on screwed nose of headstock and tighten using “C” spanner.
- Place headstock centre in the taper.
- Find centre of workpiece and drill both ends with slocomb centre drill bit.
- Attached dog (carrier) to workpiece – set cutting length ensure cutting tool will not foul on the driving pin of the catch plate.
- Set cutting tool at centre height.
- Place the workpiece between centres and tighten tailstock centre.

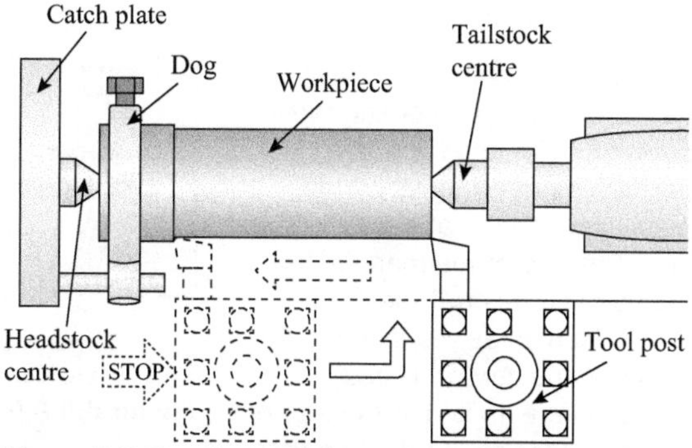

Figure 8.3 Starting spindle and touch cutting tool to diameter

- Rotate workpiece by hand.
- Set cutting speed and feed rate to suit material and surface finish required.
- Start spindle and touch cutting tool to diameter (Figure 8.3) and make a light cut to set length of cut.

- Wind tool clear of workpiece, stop spindle and check for parallelism with a micrometer (Figure 8.4).
- Adjust tailstock if taper is found.
- Determine material to be removed and rough turn diameter to marked length, leaving 0.4 mm for finishing cut.
- If necessary, turn the workpiece around placing carrier on the machined end and turn opposite end of workpiece to meet specification.

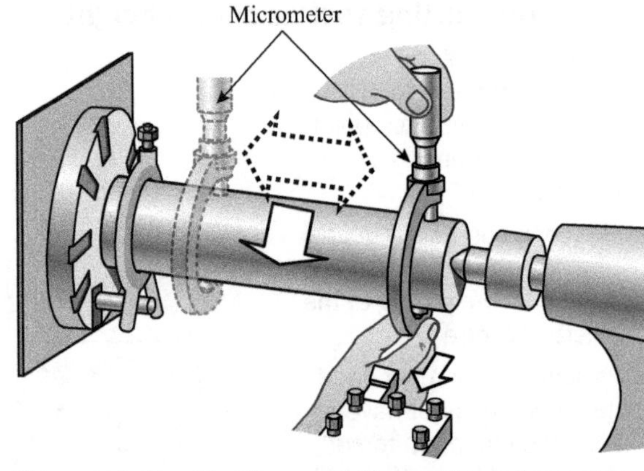

Figure 8.4 Checking for parallelism with a micrometer

Using a mandrel

A mandrel is for a workpiece that has been drilled, reamed or bored and thereby cannot be held in a chuck or on a faceplate. The workpiece is usually mounted on a mandrel and machined. The mandrel is tapered axially and pressed into the bore of the workpiece to support it between centres (see Figure 8.5). The taper is positioned so that cutting forces tend to push the workpiece towards the larger diameter of the mandrel.

There are two main types of mandrel:

- solid
- expandable.

The solid mandrel is made of hardened ground steel and tapers between 0.05 and 0.06 mm per 100 mm. It has accurately machined centres at each end. The size of the mandrel is stamped on a flat at the end, which doubles as the location to clamp the carrier. These mandrels are limited in the size of workpiece they can accommodate.

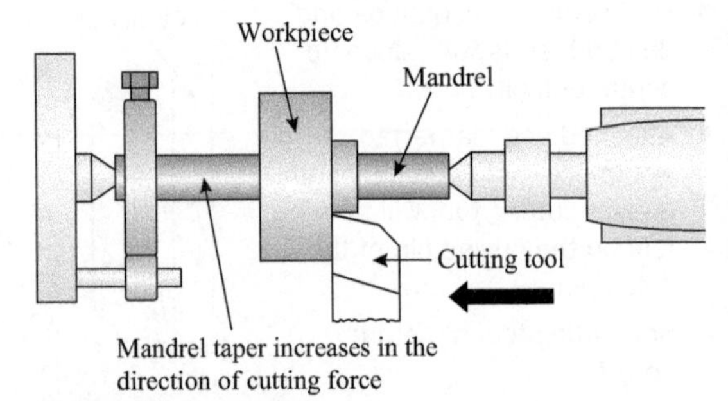

Figure 8.5 A mandrel

The expandable mandrels will accommodate a workpiece within a greater range of sizes. The expansion mandrel is arranged so that the grips can be forced outward against the interior of the bore in the workpiece.

8.2 Machining tapers

Here are some common methods for turning tapers on a lathe:

- off-setting the tail stock
- using a taper turning attachment
- using the compound slide
- using a form tool.

The method used will depend upon the length and angle of the taper. For convenience tapers can be put into two main categories:

- long tapers
- short tapers.

Machining long tapers

Most long tapers are machined using either offset tailstock or taper turning attachment.

Offset tailstock method

- Many tailstocks have a calibrated scale at the back of the tailstock that can be set using an allen screw.
- If there is no scale, a DTI will need to be used to offset the tailstock the required amount.
- Figure 8.6 shows an offset tailstock taper turning setup where the smaller diameter will be at the tailstock end.

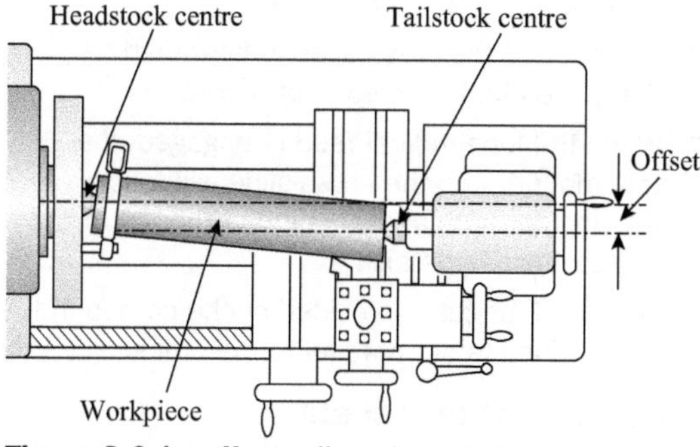

Figure 8.6 An offset tailstock taper turning setup

Calculating the taper

- Metric tapers are expressed as a ratio of 1 mm per unit length, e.g. Morse taper is 1:20.
- Small diameter d, k is unit length, l is total length of taper, L is total length of workpiece (see Figure 8.7).

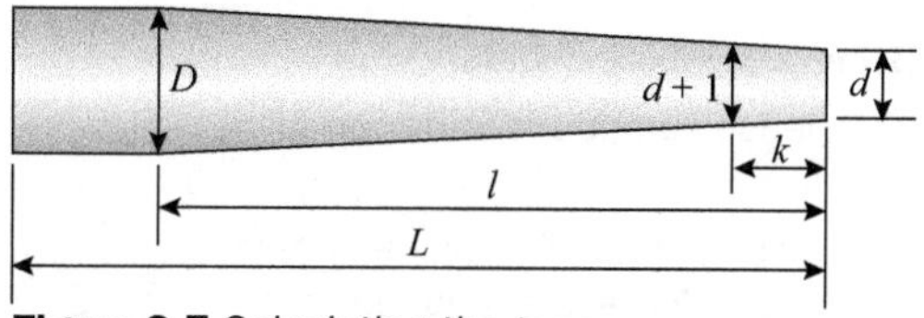

Figure 8.7 Calculating the taper

Worked example 1

Find the amount of offset required for a taper 1:25, $d = 15$mm, $l = 100$, $L = 125$, $k = 25$

$$D = d + \frac{l}{k} = 15 + \frac{100}{25} = 15 + 4 = 19 \text{ mm}$$

- Amount of offset required $= \frac{D - d}{2 \times l} \times L = \frac{19 - 15}{2 \times 100} \times 125 = \frac{4}{200} \times 125 = 2.5$ mm

Note: If the tailstock does not have a calibrated scale for offsetting the tailstock centre, a DTI can be used to set the offset amount required.

- Large diameter $D = d$ plus total amount of taper.
- Amount of taper/unit length $(k) = (d + 1) - (d) = 1$ mm
- Therefore, the amount of taper $= \frac{1}{k}$
- Therefore, total amount of taper $= \frac{1}{k} \times l$ or $\frac{l}{k}$; $D = d + \frac{l}{k}$

The taper turning attachment

- The taper turning attachment (Figure 8.8) is fitted to the back of the lathe bed.
- The guide bar is pivoted at the centre and is graduated in degrees; it can be pivoted on either side of the zero graduation and set longitudinally at the desired angle to the lathe axis.
- The cross-slide is made free from its screw by removing the binder screw.
- The back of the cross-slide is tightened to the guide block by means of a bolt.
- When the longitudinal feed is engaged, the tool mounted on the cross-slide will follow the angular path as the guide bar is set at an angle of the lathe axis.
- The depth of cut is provided by the compound slide, which is set parallel to the cross-slide.

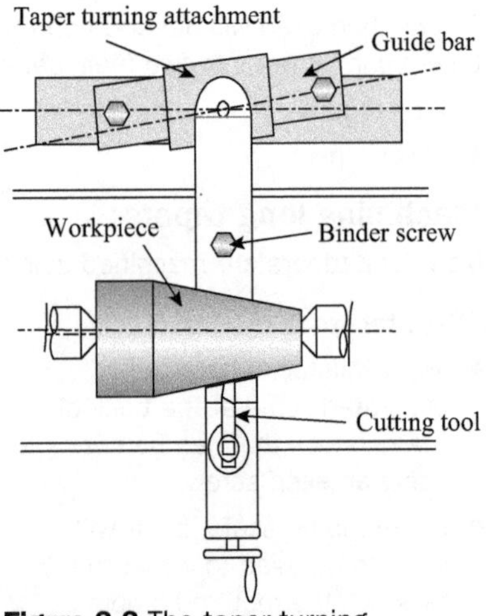

Figure 8.8 The taper turning attachment

Machining short tapers

Most short tapers are machined using either compound slide or forming tool.

Using a compound slide

- Release compound slide clamping screw.
- Set compound slide angle at half the included angle required (see Figure 8.9).
- Tighten compound slide clamping screw.
- Set cutting tool on centre height. **Note:** The angle will not be accurate if the tool is above or below the centre height.
- Calculate and set cutting speed.
- The depth of cut will be determined by turning the cross-slide clockwise.
- The cut will be made by turning the compound slide clockwise and taken off by turning the same wheel anti-clockwise.

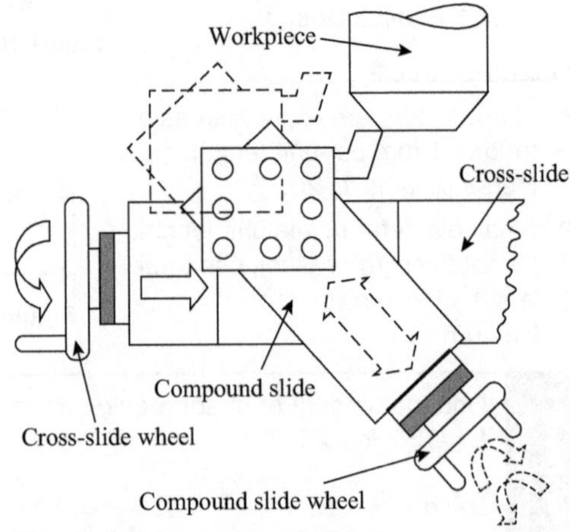

Figure 8.9 A compound slide

- With the cutting tool clear of the end of the workpiece, a further cut is applied and the cross-side turned and unwound after the cut.
- This is repeated until the length of taper is achieved.

Using a forming tool

- The cutting tool is ground with a straight cutting edge.
- The angle of the cutting edge is ground at half the included angle as the required taper.
- The cut is applied by turning the cross-slide which is perpendicular to the axis of the lathe.
- The taper length must be less than the width of the tool.
- Normally the length of taper is achieved by producing the small diameter of the taper.

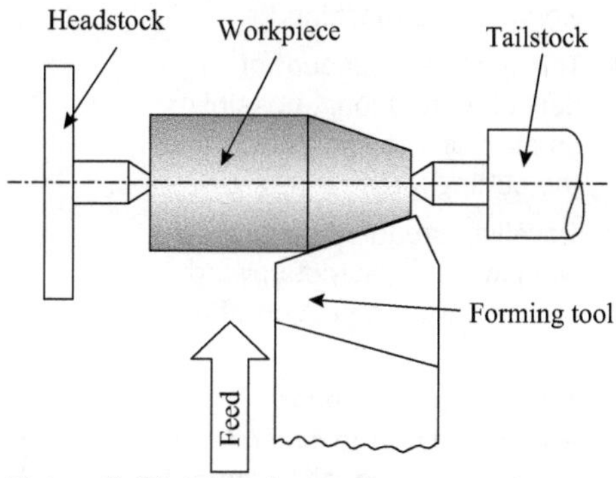

Figure 8.10 A forming tool

Internal taper turning

- Generally internal tapers produced on a lathe are short tapers.
- Longer internal tapers are mainly produced on other machines by different methods.
- Predominately compound slide or forming tools are used to machine internal tapers.
- Figure 8.11 shows an internal taper being machined using a compound slide.

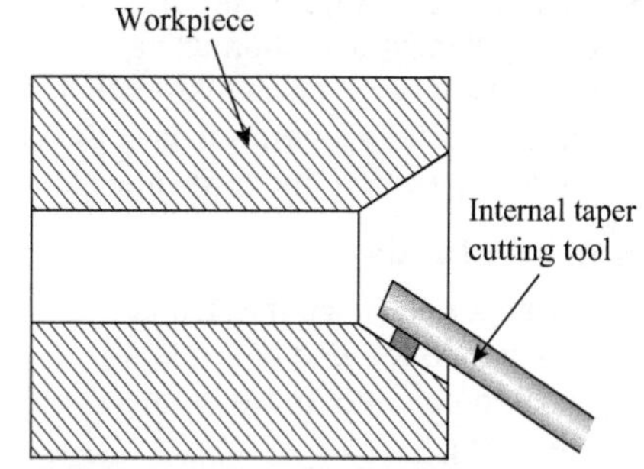

Figure 8.11 An internal taper

8.3 Machining using fixed and travelling steadies

Steadies are a lathe accessory used to support cylindrical workpieces that require support, because they are long in length and small in diameter or are unable to be supported by a tailstock centre because of the machining operations being carried out.

Depending on the machining operations one of two main types of steady will be used:

- a travelling steady
- a fixed steady.

A travelling steady

- A travelling steady (Figure 8.12) is attached to the saddle of the lathe and travels with it as the workpiece is machined.
- This limits the amount of deflection of a long, possibly small, diameter workpiece under the cutting forces that are applied.
- Travelling steadies usually have two jaws that restrict upward and backward defection of the workpiece.
- The jaws are curved to accommodate the circumference of the workpiece, although some have a running wheel that turns with and supports the workpiece.
- There are two ways to set up the machining operation using a travelling steady: having the cutting tool **trailing** the steady's support of the workpiece on its original diameter; having the cutting tool in **advance** of the steady's support of the workpiece on the newly machined diameter.

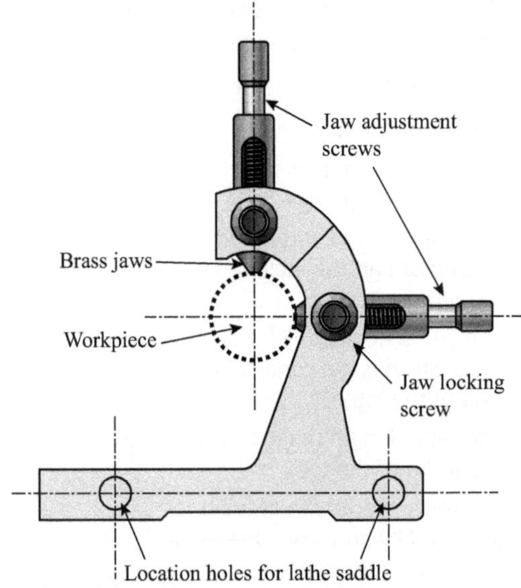

Figure 8.12 A travelling steady

A fixed steady

- A fixed steady (Figure 8.13) is attached to the bed of the lathe bed and does not move during machining.
- This type of steady typically has three jaws that support the workpiece.
- The steady is located on the headstock end of the lathe bed so that the saddle and tool post can be used.
- It allows machining of the end of the workpiece without a centre.
- Fixed steadies can accommodate larger diameter workpieces that require boring or internal machining at a distance from the headstock of the lathe.
- The workpiece may be set up using a tailstock centre which is then removed to allow the required machining operations.

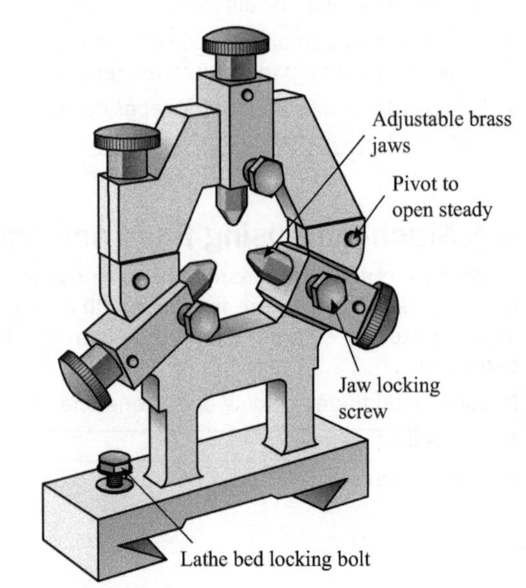

Figure 8.13 A fixed steady

8.4 Calculating and selecting simple and compound gear trains

Sometimes it is necessary to mesh two or more gears with each other to transmit power and direction from one shaft to another. Such a combination is called a **gear train**, and the type of train used depends upon the velocity ratio required and the relative position of the axes of shafts.

Types of gear train

There are two main types of gear train:

- a simple gear train
- a compound gear train.

Simple gear trains

- Simple gear trains have only one gear on each shaft (see Figure 8.14).

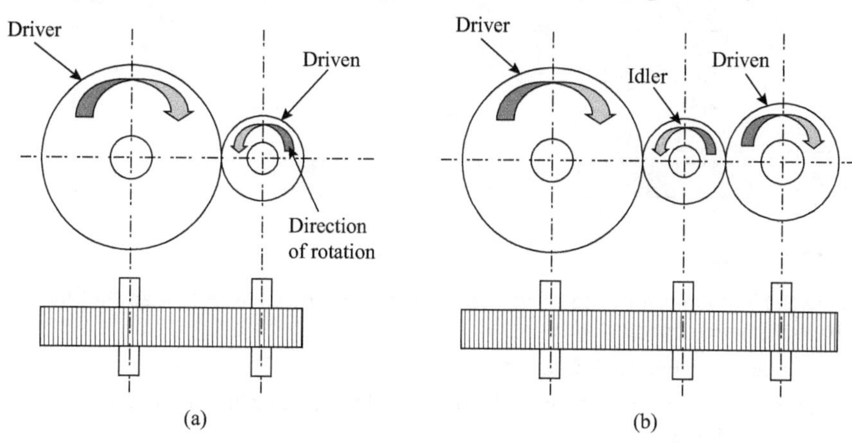

Figure 8.14 Simple gear trains

Key terms

- The gear on the shaft that supplies power to the gear train is called the **driver**.
- The gear on a shaft that is turned by the driver is called the **driven** (sometimes the **follower**).
- A gear positioned in between the driver and driven gears is called an **idler** or **intermittent gear**.
- The last gear in a train is called the **output gear**.

- As power is transmitted from one gear to the next, each gear turns in the opposite direction.
- The gear ratio (or velocity ratio) is found by the number of teeth on each gear. The equation is: $\text{Gear ratio} = \dfrac{\text{no. teeth on driven}}{\text{no. teeth on driver}}$

- If the driver has 60 teeth and the driven has 30 teeth, the ratio is 1:2.
- The RPM for each gear will be at the same ratio, see Table 8.1.
  Note: The number of teeth on an idler gear does not affect this ratio.

Table 8.1

Driver	Driven
60 teeth	30 teeth
100 RPM	200 RPM

Compound gear trains

- A compound gear train is where there is more than one gear on at least one shaft.
- Gear A is the driver.

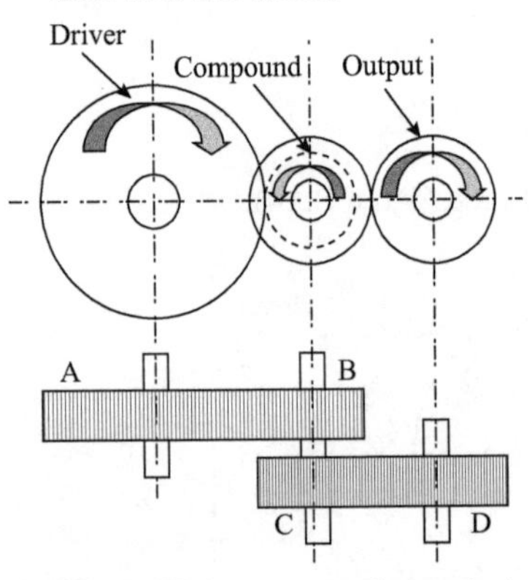

Figure 8.15 A compound gear train

- In Figure 8.15 the intermediate shaft has two gears (B and C) fixed to it so that they may have the same speed (RPM).
- Gear B is a driven, while Gear C is a driver.
- One of these two gears meshes with the driver and the other with the output (driven) gear D, fixed to the next shaft.

Worked example 2

Calculate the compound gear train velocity ratio:

Gear A	Gear B	Gear C	Gear D
120 teeth	40 teeth	60 teeth	40 teeth

$$\frac{\text{Driven (B)}}{\text{Driver (A)}} = \frac{40}{120} = \frac{1}{3}$$

$$\frac{\text{Driven (D)}}{\text{Driver (C)}} = \frac{40}{60} = \frac{2}{3}$$

The velocity ratio of the compound gear train is $\frac{1}{3} \times \frac{2}{3} = \frac{2}{9}$

If Gear A rotates at 30 RPM, the RPM of Gear D will be $30 \times \frac{9}{2} = 135$ RPM

1. Figure 8.16 shows a taper to be machined using the offset tailstock method. The taper ratio is 1:20, $d = 20$ mm, $l = 150$ mm, $L = 175$ mm, $k = 30$ mm.

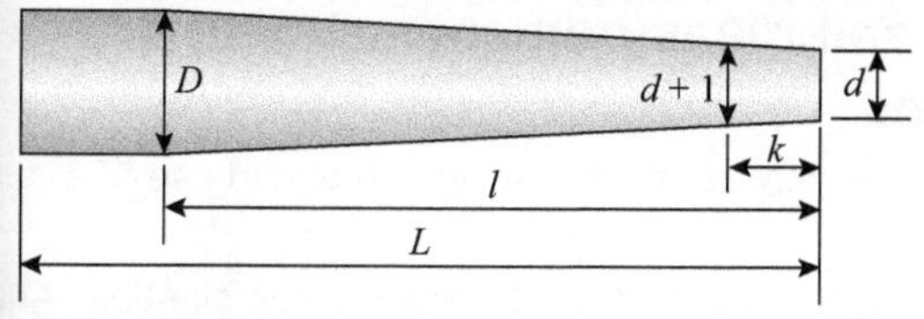

Figure 8.16

a) Calculate the amount of offset required to machine this taper.

b) State whether the offset will be towards or away from the lathe operator.

2.

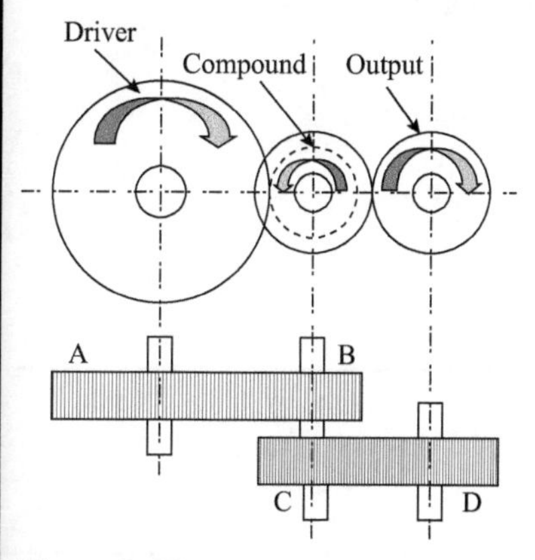

Figure 8.17

Figure 8.17 shows a compound gear train.

The gears used are shown in the table.

If Gear A rotates at 40 RPM, calculate the RPM of the output Gear D.

Gear A	Gear B	Gear C	Gear D
100 teeth	50 teeth	80 teeth	40 teeth

Work with a classmate to review what went well, and any issues that arose during the activity saying how you overcame them.

Put a copy of your work in your portfolio of evidence.

Why don't you?

9. Performing sheet metal operations

Option B Section 3.6.1

9.1 Accuracy of layout and measurement for development patterns

Note: You need to be familiar with the location, the name, and intended use of a range of hand tools including:

- **Marking out tools** – scribes, straight edge, dividers, try square, combination square, protractor, steel rule, trammels, circumference ruler and tape rule.
- **Cutting tools** – straight, curved and universal snips, hacksaws and files.
- **Hammers and mallets** – stretching and planishing hammer, and boxwood and rubber mallets.
- **Hand jointing tools** – rivet snaps and pop riveting guns.

Marking out procedure

Before starting, it is important to ensure the material:

- is correct for the required job
- has straight or square edges
- is not warped or twisted
- is not damaged, and has no surface defects.

Did you know?

A **circumference rule** is used to measure the diameter, and it gives the equivalent circumference without the need of a calculation.

Marking out

- Layout dye: use water paint for rough surfaces, and marking blue dye out for smooth surfaces.
- Datum lines are fixed or starting points, line or edges. They make marking out more accurate if all measurements are from the appropriate datum point.
- Centre lines are scribed from the datum lines or edges to establish the positions of holes, slots, radii and other details.
- Outlines show the dimensions of the work piece and indicate the location and amount of metal to be removed, folded or bent which determine the finished shape.
- Often the marking includes allowances for folding, bending and jointing.
- At this stage the marking out is checked for accuracy, and any adjustments are made.
- Establish marking out lines by following checking of dimensions. Witness marks can be added to permanently indicate the position of the outlines. These are light, uniform indents made with a punch. The witness marks ensure the marking out will not be lost among the scratches that occur during cutting, forming and jointing.

Accuracy of marking out

- Accuracy could be defined as the maximum permissible error in any particular measurement or marking out.
- Error in measurement and marking out is generally accepted as one-half the magnitude of the smallest calibration on the measurement scale.
- Because measurement and marking out requires the alignment of both ends of the rule or measuring method from the starting point of the measurement to the end point of measurement, the error could be twice that equalling the smallest scale (calibration) reading.
- The principle cause of the error is **error of parallax**. For accurate measurement, the measurer's eye must be aligned directly above the mark being read. This is to avoid parallax errors which will give rise to inaccurate measurement when viewed from an angle.

Layout

Sheet metal layout (marking out) is done while the material is in flat sheet form. It then may be bent, folded and joined to itself or other sheets to form the required component shape. To the untrained eye the layout may bear no resemblance to the finished component. The layout shape produced in the flat is called the **development.**

Developments

It is important you have a mental image of what the general development is for a range of common shapes. Figure 9.1 demonstrates the nature of how the development in the flat (2D) is folded into 3D. Important considerations when producing the development are allowances for other subsequent operations such as wired edges and jointing (see Figure 9.2). Jointing allowances may vary depending on the method of jointing to be used.

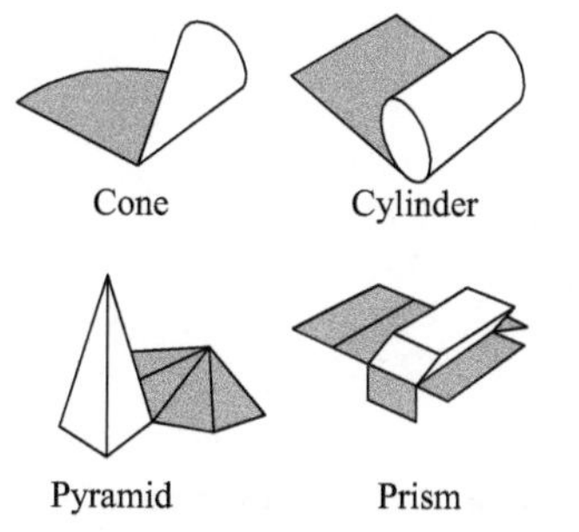

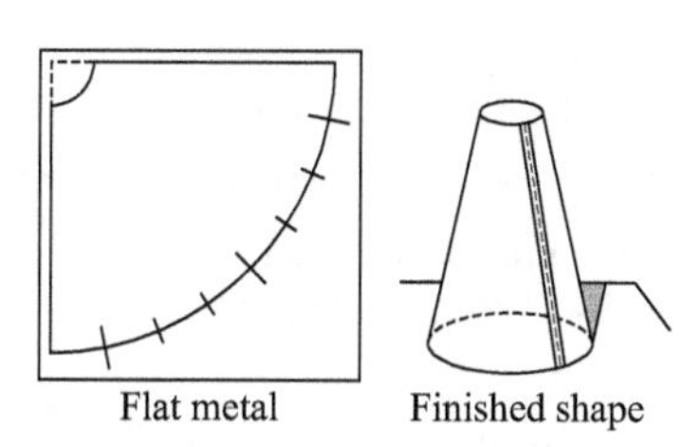

Figure 9.1 Folding from 2D to 3D

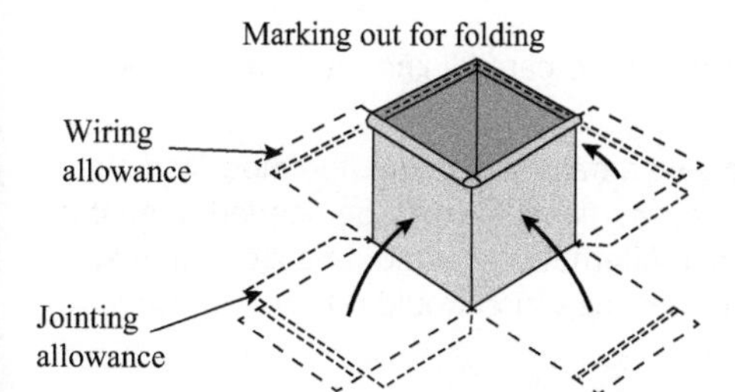

Figure 9.2 Jointing and wiring allowances

9.2 Principles of pattern development for making simple templates

There are three main methods of producing developments or layouts:

- radial line
- parallel line
- triangulation.

The radial line method

An example of developing a cone using a radial line is shown in Figure 9.3.

1. Obtain work drawing and specifications.
2. Select and collect layout tools: flat steel rule, straight edge, dividers, trammels, prick (centre) punch and scriber.
3. Select sheet of correct material and size for task.
4. Clear marking out table and wipe with cloth to ensure cleanliness.
5. Draw the side elevation GHKA with extended centre line.
6. Extend lines GH and and AK to O.
7. Draw arc AG (radius of cone base) and divide into six equal parts: A, B, C, D, E, F, and G.
8. At centre O, draw the arcs AX and KY ensuring they intersect the centreline DX.
9. Using dividers, take the distance *z* (=AB) and step off 12 times along the arc AX as shown.
10. Number each step: A^1, B^1, C^1, D^1, E^1, F^1, G^1, F^2, E^2, D^2, C^2, B^2, and A^2.
11. Extend A^1, B^1... B^2, A^2 to centre O.
12. The development for the right cone is A^1–A^2– L^1– K^1 plus the joining allowance.

This profile is then cut out.

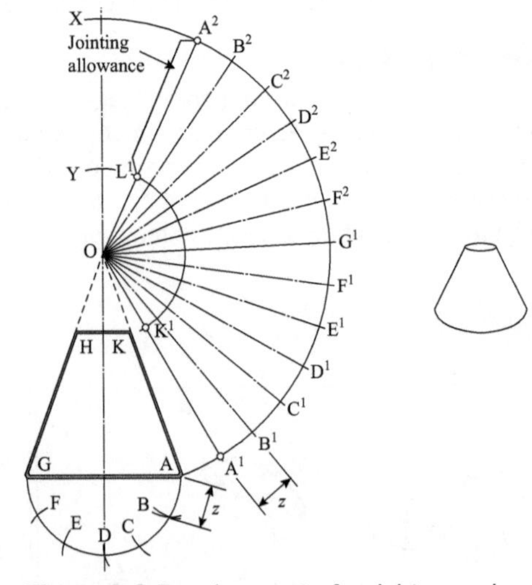

Figure 9.3 Development of a right cone by a radial line

The parallel line method

An example of developing a truncated cylinder using the parallel line method is shown in Figure 9.4.

These components are often made in pairs to give a bend in ducting, the most common being a 45° angle to give a 90° bend. In these cases flanges would be needed to enable the two truncated cylinders to be joined together. Alternatively, without flanges another jointing process would be needed, for example, welding which would rely on the accurate location of the cylinders during welding.

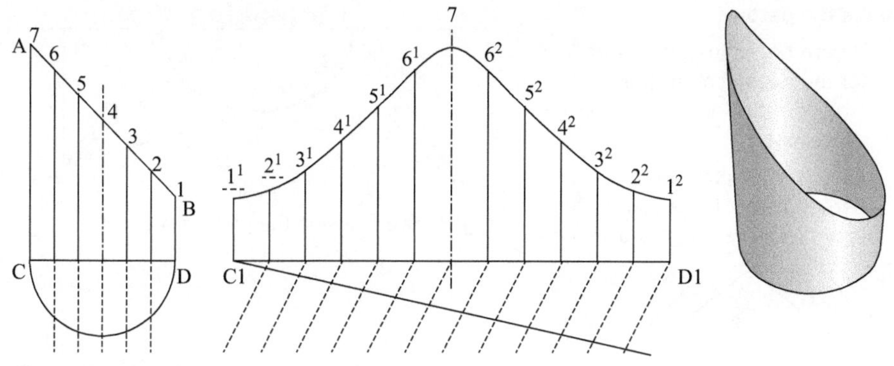

Figure 9.4 Developing a truncated cylinder using the parallel line method

1. Obtain working drawing and specifications.
2. Select and collect layout tools: flat steel rule, straight edge, dividers, trammels, prick (centre) punch and scriber.
3. Select a sheet of the correct material and size for task.
4. Clear the marking out table and wipe with a cloth to ensure cleanliness.
5. Draw side elevation (A, B, C, and D) and half plan view (semicircle) at the centreline. Divide the semicircle circumference into six equal parts.
6. Extend lines from the semicircle circumference through CD to intersect line AB.
7. Draw datum line (C1 to D1) equal to the circumference (πD) of the cylinder and divide into 12 equal parts plus centreline and extend upwards.
8. Project intersections at AB to meet divisions extended from the circumference datum (1 to 1^1 and 1^2, 2 to 2^1 and 2^2... 6 to 6^1 and 6^2 and 7 to 7 (centreline).
9. Mark intersection and join intersections with smooth line to form the development.

The triangulation method

An example of a pattern development for making simple templates using triangulation is shown in Figure 9.5.

Transition pieces are normally made to connect two different sections, such as square to round (as shown), rectangular to round, and round to round. These sections may be on or off centre. Triangulation is more time consuming and more difficult than parallel line or radial line developments.

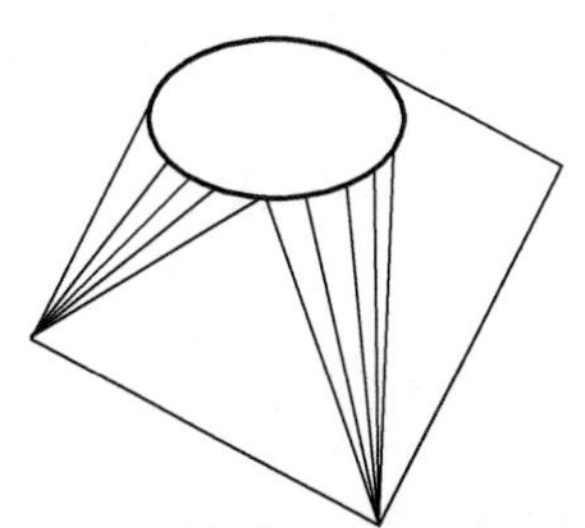

Figure 9.5 Making simple templates using triangulation

To make the pattern

1. Before beginning the layout for the development the "true lengths" lines must be determined.
2. Each true length line is perpendicular to its plane which is done by drawing the plan view.

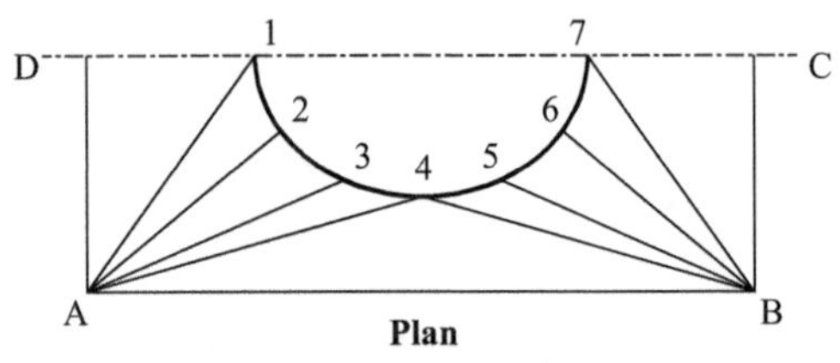

Figure 9.6 Drawing the plan view

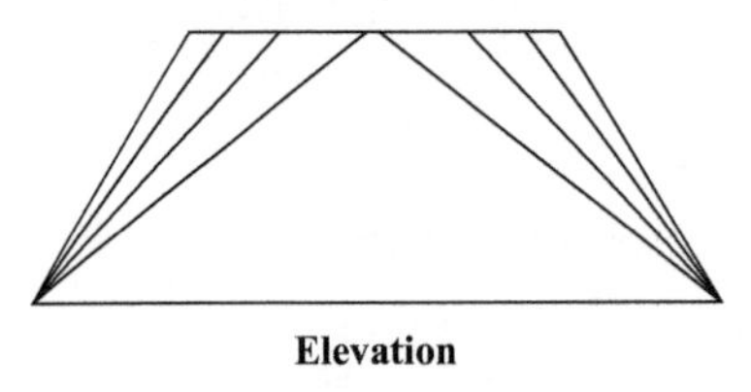

Figure 9.7

3. Using dividers transfer from the plan view produce a true length bar.

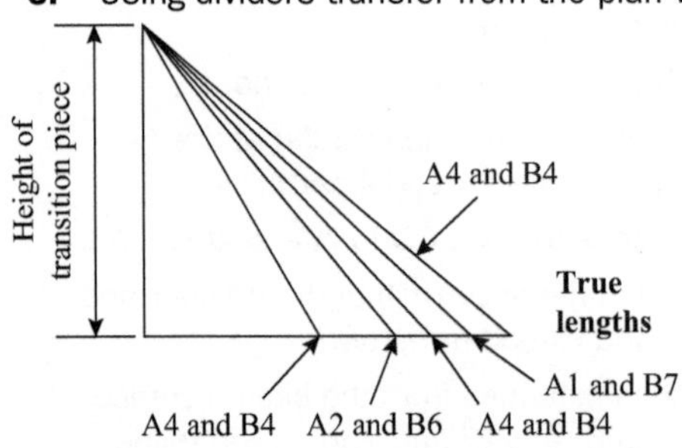

Figure 9.8

4. Draw true line AB from plan view.

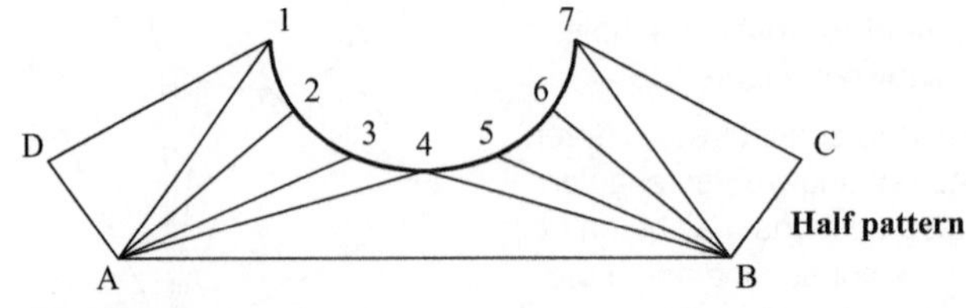

Figure 9.9

5. Next, locate point 4 by swinging the true length of A4 from A, and by swinging the true length of B4 from B.
6. Next point 4 is located, then find point 3. This is done by swinging the true length of A3 from A and by swinging the distance 4–3 from 4 from the plan view.
7. Point 2 is found next and then point 1. After this, the other side of the pattern is laid out by locating points 5, 6, and 7, in that order.

8. The curve of the pattern is drawn through these points. This can be done freehand or by using a flexible curve which will be more accurate.
9. After the curve points 1 to 7 are located, then points C and D must be found to complete the half pattern. To find C, take the distance BC from the plan view and swing it from B on the pattern.
10. Take the true length of C–7 and swing it from 7. The intersection of these two arcs gives us point C, and the lines 7C and BC can be drawn in. Point D is located in the same way.
11. Take AD from the plan view and swing it from point A. Then take the true length of D–1 and swing it from 1. This locates D and the lines can be drawn.

Notes:

- It is important to check accuracy at each stage.
- Two of the half patterns are required to produce a transition piece and, because they are symmetrical, the first half pattern can be used as a template for the second.

9.3 Sheet metal cutting tools

Sheet metal cutting tools need to be used safely, efficiently, and accurately.

Safety will be determined by the training and experience of the operator. It is important that you know how to use each piece of equipment. If you have not been shown or are not sure, ask for guidance.

Efficiency will be a function of selecting, caring for, and using the correct equipment for the task.

The quality of the finished component will depend on the **accuracy** of cutting, and this will be dependent on the accuracy of marking out.

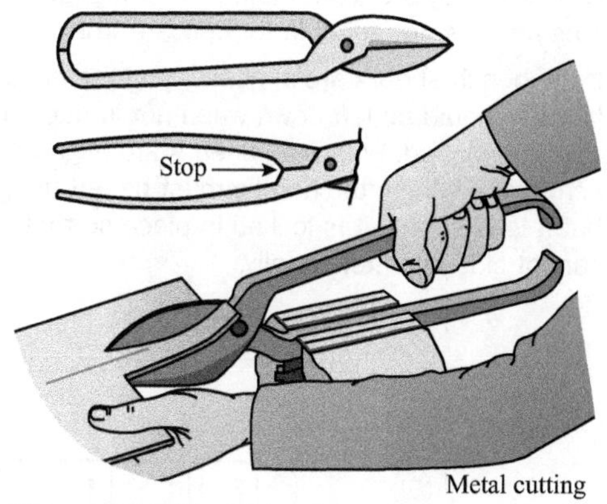

Figure 9.10 Straight snips

Straight snips (Figure 9.10) are for general purpose, cutting and trimming where free access and movement of the handles is obtained.

Greater cutting forces can be generated by securly clamping one of the handles in a vice and using them like bench shears.

Curved snips (Figure 9.11) are for circles and curved shapes, varying in size from 20–35 cm in steps of 5 cm.

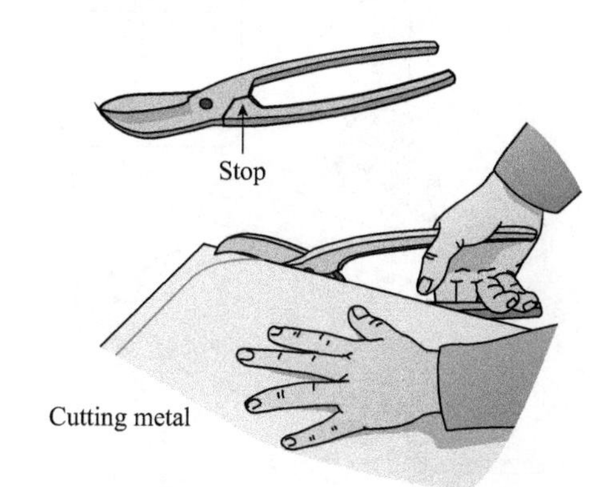

Figure 9.11 Curved snips

Universal snips (Figure 9.12) are for cutting enclosed shapes in sheet metal. They are available in left-handed and right-handed, which is determined by the position of the top blade to the bottom one. Some operations will require both hands.

When using snips and shears (Figure 9.13) the handles should be held as far from the fulcrum as possible to maximise cutting forces. The blade should be at right angles to the workpiece aligned to the layout and should never be completely closed, as this will distort the metal and give a ragged edge. Use finger control where stops are not fitted to prevent trapping the palm.

Bench shears (Figure 9.15) are used for cutting thicker gauge sheet and have a stated capacity. They provide a quick and safe way of cutting thicker material, but must not be used beyond their stated capacity. They are used mainly for straightforward cuts and are not suitable for delicate work.

If the bench shears are aligned *along* the bench, the handle should be left down when not in use so the jaws are closed. When the shears are aligned *across* the bench, the handle must be left upright but it is essential it is locked in place so that it cannot close unintentionally.

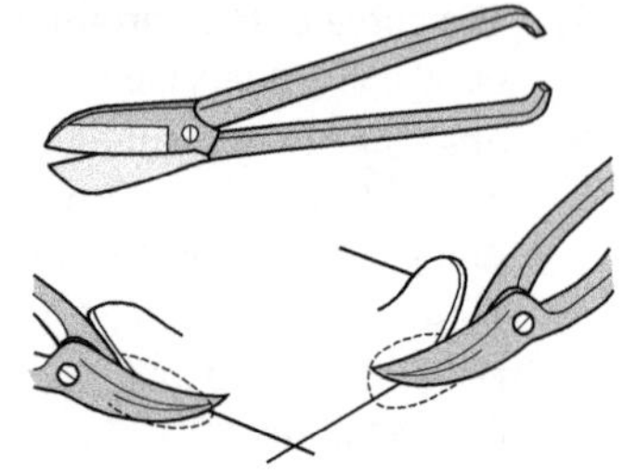

Figure 9.12 Universal snips

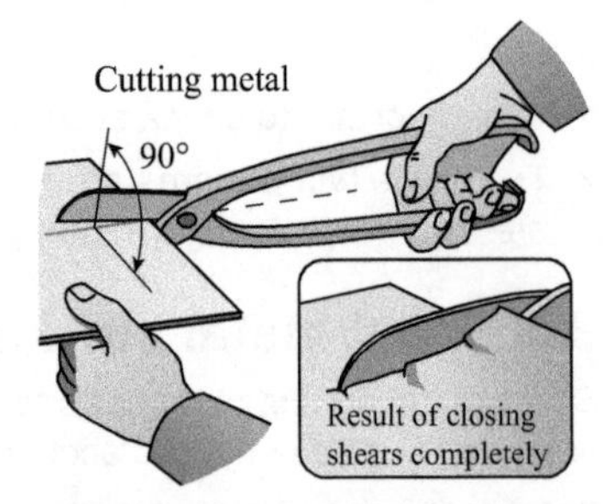

Figure 9.13 Using hand shears

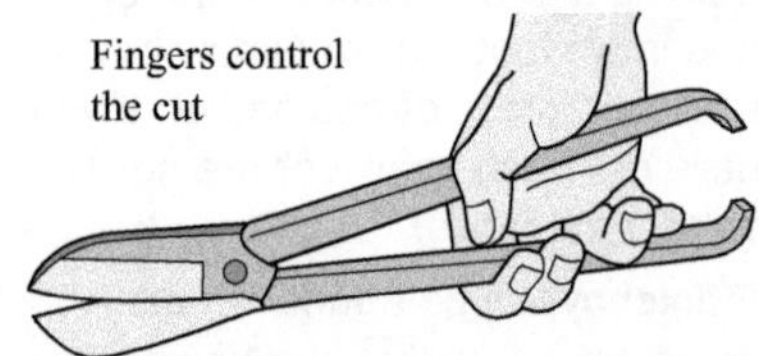

Figure 9.14 Fingers control the cut

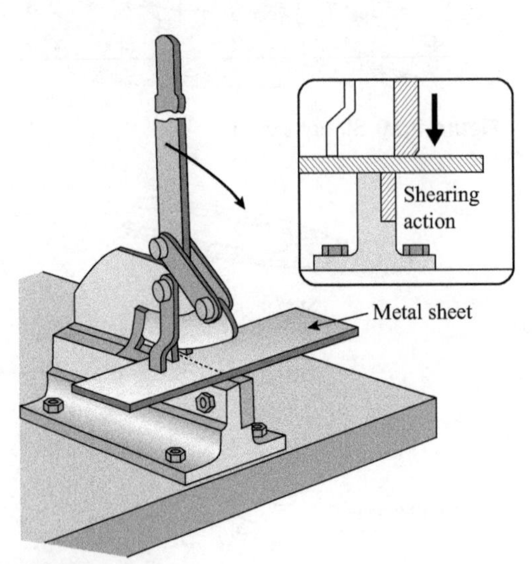

Figure 9.15 Bench shears

In the treadle guillotine (Figure 9.16), the bottom blade is fixed to the bed of the machine and the treadle operates the top blade. The material to be cut is placed on the machine bed and held securely in place by hand. A hold down clamp is operated when the treadle is depressed and providers cover for the cutting blade.

Safety: Ensure guards are in place before using this guillotine.

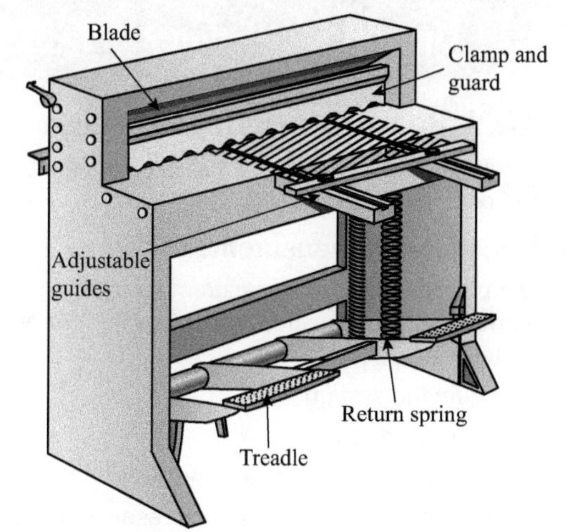

Figure 9.16 A treadle guillotine

Why don't you?

Produce a truncated cylinder (Figure 9.17) using the **parallel line** method.

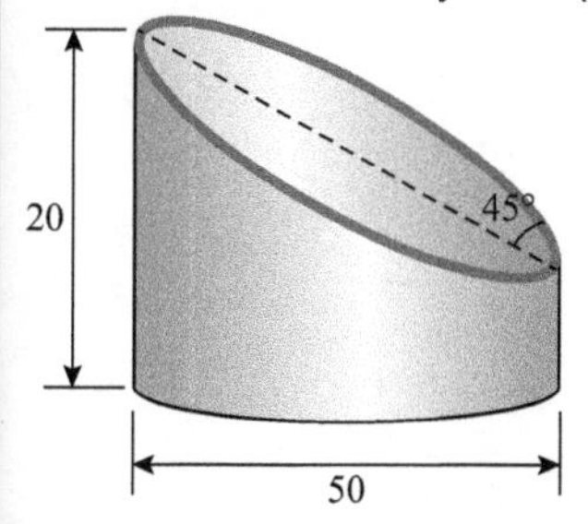

Figure 9.17

This can be done by using the inside of a cardboard cereal box.

Guidance: Draw the side elevation of the truncated cylinder with a semicircle with the diameter of the cylinder drawn on the centreline on the base of the cylinder. Divide the semicircle into six equal parts and project upwards until they meet the slope of 45°. Number these 1–6. Extend the base line of the cylinder to the right and mark off the length of the cylinder circumference ($\pi\Delta = 3.142 \times 20$). Divide this line into 12 equal parts, and project upwards higher than the top of the cone. Number these 1–12 also. Project each intersection on the 45° slope across to the right until each intersects the appropriate line 1 to 1^1 and 1^2, 2 to 2^1 and 2^2...5 to 5^1 and 5^2, 6 to 6^1 and 6^2, and 7 to 7. Joint these intersections with a smooth curved line. Add a small tab at the short right-hand side (jointing allowance). This will form the **development**. Cut it out and construct the truncated cylinder by bringing the two short sides together.

Work with a classmate to agree what went well and what could have been improved. Put the truncated cylinder in your portfolio of evidence along with any notes.

10.1 Reading and measuring

In this unit we will look at three measuring instruments: the micrometer, the vernier calliper, and the surface plate.

Principles of measurement

The principle of linear measurement is one of comparing the workpiece being measured, marked out, or inspected against a known standard. The two recognised international systems are:

- the metric system
- the imperial system.

In this section we will deal with the metric system only.

Working standards are calibrated against master standards to ensure uniformity and consistency essential for the concept of interchangeability.

The micrometer

The different parts of a micrometer are shown in Figure 10.1.

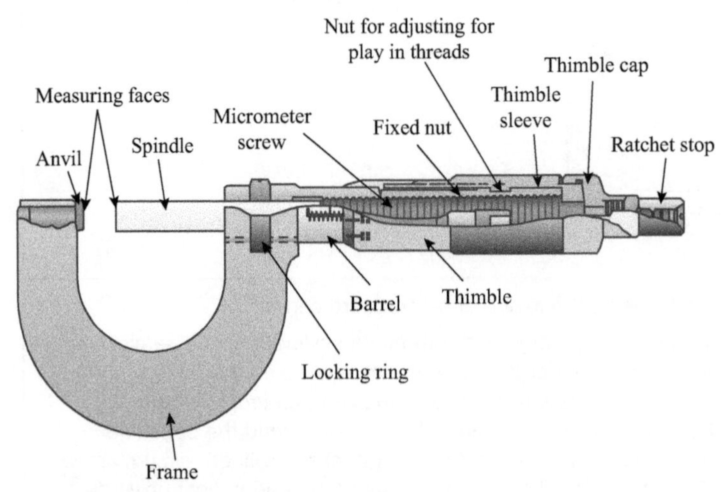

Figure 10.1 A 0–25 metric micrometer

- The pitch of the micrometer screw is $\frac{1}{2}$mm, so the spindle will advance or retract 0.5 mm for one **complete revolution** of the thimble.
- The thimble has a bevelled edge and is divided into 50 graduations.
- Each graduation represents $\frac{1}{50}$ of $\frac{1}{2}$mm which equals $\frac{1}{100}$mm (0.01 mm).
- The barrel is graduated in whole millimetres above the axis line and $\frac{1}{2}$mm below it.

Reading a metric micrometer

1. Read the number of whole millimetres above the axis line visible on the thimble.
2. Add any $\frac{1}{2}$mm below the axis line on the thimble.

3. Read the number of millimetres on the thimble that coincides with the axis line on the thimble.

For example, see Figure 10.2:

Whole millimetres above axis line:

10 = 10.00 mm

Additional $\frac{1}{2}$ mm below axis line:

1 = 0.5 mm

Number of millimetres coinciding with axis line:

22 = 0.22 mm

Total reading = **10.72 mm**

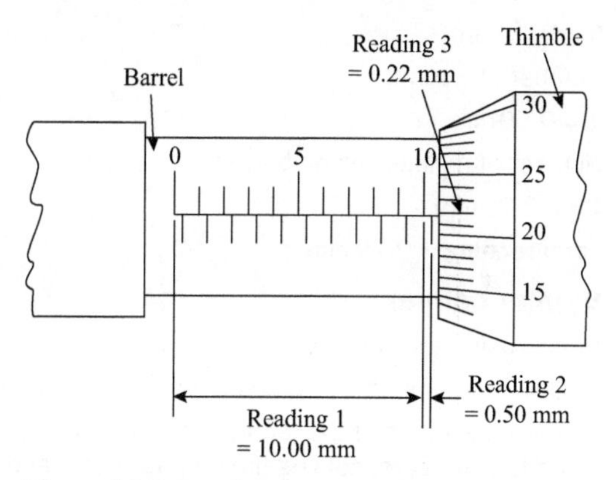

Figure 10.2 Reading a micrometer

Using a micrometer

- Metric micrometers are named by the size, which normally goes up in 25 mm stages e.g. 0–25, 25–50, 50–75, and so on.
- Small micrometers can be operated by one hand, freeing the other hand to hold the workpiece.
- Hold the frame of the micrometer in the palm of the hand and gripped by the second, third and fourth fingers.
- The first finger and thumb are used to rotate the thimble.
- The final adjustment is done by turning the ratchet with the forefinger and thumb. This ensures uniform pressure and consistency of reading.
- Some micrometers are not fitted with a ratchet, so the final adjustment is made with the thimble. This ensures the right amount of pressure is used to avoid a faulty reading and excessive wear to the micrometer screw and measuring faces.

Depth micrometer

Depth micrometers are used to measure the depth of holes and shoulders. Instead of the two measuring faces seen in the micrometer in Figure 10.1, the depth micrometer has a flat head and a spindle. The other important difference is that the scale is reversed.

Using a depth micrometer

- Turn the thimble so the face of the spindle clears the bottom of the hole.
- Place the micrometer head on the workpiece over the hole to be measured, where there is a suitable seating surface for the micrometer head.
- Turn the thimble until the face of the spindle touches the bottom of the hole.
- Read off the measurement on the micrometer.

Note: the scale is read in **reverse**.

For example, see Figure 10.3:

Number of whole millimetres:

9 = 9.00 mm

Number of $\frac{1}{2}$ mm:

1 = 0.50 mm

Number of millimetres on barrel:

29 = 0.29 mm

Total reading = **9.79 mm**

Figure 10.3 Reading a depth micrometer

Vernier caliper

Before using a vernier caliper:

- Clean measuring instruments before and after use.
- The parallelism of the jaws can be checked by placing the jaws together, without undue pressure, holding them up to a light and viewing any gap.
- For internal measurement the nibs must be examined for damage or wear. An ordinary micrometer can be used to check and any allowances made.

Using a vernier caliper

- Check that both clamp screws are released.
- Hold the vernier in one hand, slide the moveable jaw with the thumb until the jaw makes contact with the workpiece.
- With the other hand, lock the clamping block with the screw – this is the block furthest from the workpiece.
- Using the fine adjustment wheel, adjust the moveable jaw to just grip the workpiece without undue pressure.
- Turn the screw of the vernier block to clamp it in place.
- Read the vernier in good light.

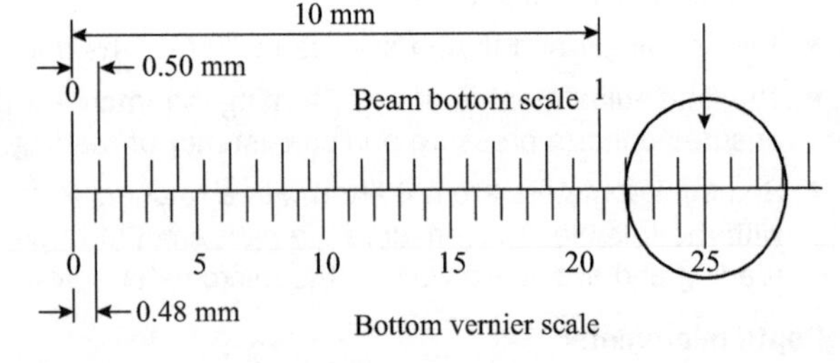

Figure 10.4 The vernier calliper

Note: Most vernier calipers have both metric and imperial calibrations. The metric scale is usually the bottom scale on both the (fixed jaw) beam and vernier (moveable jaw) block.

There are two scales to consider: the beam bottom scale engraved on the beam on the instrument, and the bottom vernier scale on the moveable vernier block.

- On the beam scale 0 to 1 is 10 mm, which is divided into 10 parts of 1 mm. Each millimetre is divided in two, $\frac{1}{2}$ mm.
- On the bottom vernier, 12 mm is divided into 25 parts (see coinciding lines on Figure 10.4) $\frac{12}{25} = 0.48$ mm.
- Every fifth division is numbered so the difference in length between one small division on the beam bottom scale and one bottom vernier scale is $\frac{1}{2}$ mm. See Figure 10.4 $0.50 - 0.48 = 0.02$ mm

Reading a metric vernier caliper

For example, see Figure 10.5:

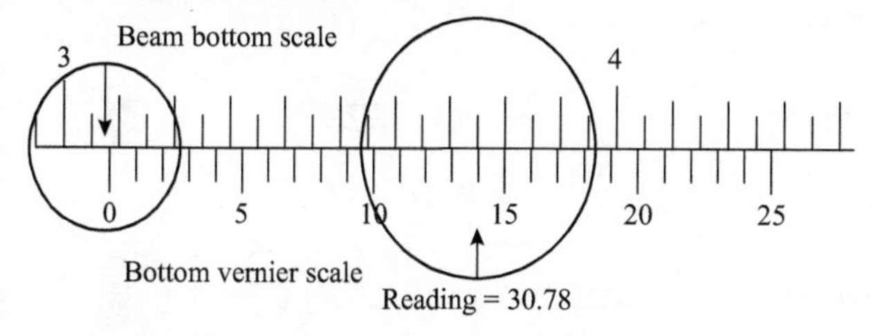

Figure 10.5 Reading a metric vernier caliper

- To the left (small circle) the 0 on the bottom vernier scale is $\mathbf{3\frac{1}{20}}$ = 30.50 mm
- The coinciding line (large circle) on the vernier scale with the beam scale is **14** = 14 × 0.02 = 0.28 **Total reading** = **30.78 mm**

Vernier height gauge

- This instrument has a column or beam engraved with a calibrated metric scale that is attached to a base.
- On the column are two blocks; one is a block that can be clamped to the column (the locking block), while the other has a vernier scale that is attached to the locking block via a fine adjustment screw.
- The vernier block is engraved with a vernier scale that can align with the metric scale on the column.
- Attached to the vernier block is a sharp scribing finger that is used to make accurate marking out (layout) lines on components.
- The scale registers height from the precision surface, so the bottom of the base must be clean and free of any damage otherwise the accuracy of the instrument will be compromised. For this reason it is often used for comparing differences in height.
- When not in use the height gauge must be returned to storage in a protective box.

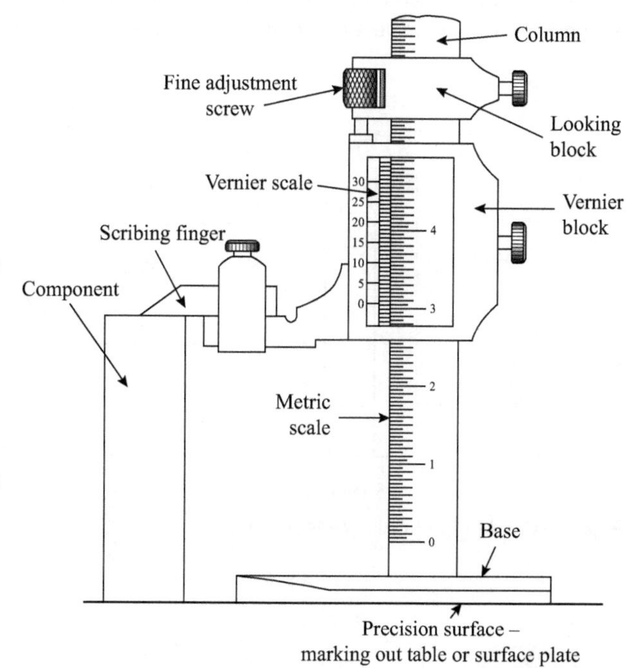

Figure 10.6 A vernier height guage

Checking the accuracy of a vernier height gauge

- The height gauge must be on a clean surface table.
- Wring the slips together and place them on the surface table.
- Lower the height gauge scribing finger onto the wrung slips.
- The reading on the vernier should equal the height of the slips.

Note: The method of reading the height gauge vernier is exactly the same as for the vernier calipers.

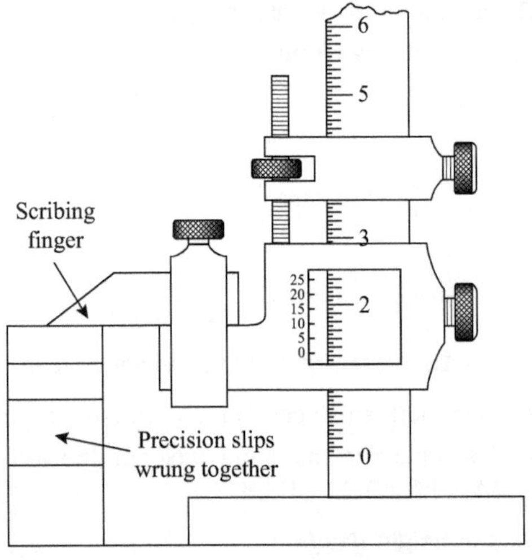

Figure 10.7 Checking the accuracy of a guage

10.2 Measuring and laying or marking out

Before starting to mark out it is important to establish two faces that are flat, straight, and square with each other, marking them as a face side and face edge. Setting these data (plural of *datum*) is a vital part of the making process. All measurements are made from these datum faces. One convention of identifying datum faces is to mark the first datum edge with a single slanted line (/) and the second with double lines (//).

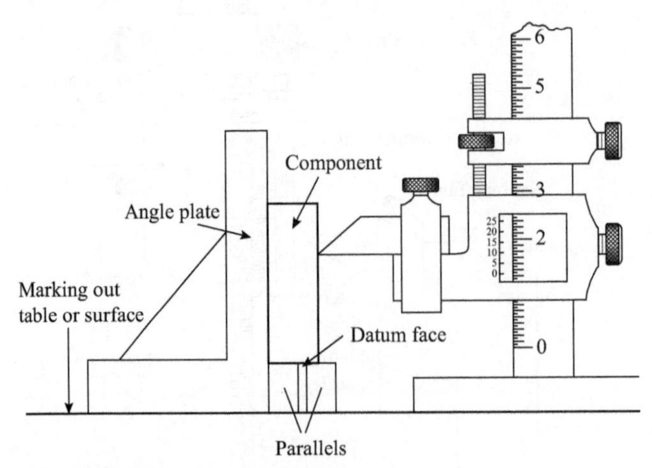

Figure 10.8 Marking out arrangements

- The accurate marking out of engineering components requires a flat marking out table or **surface plate**.
- The component is sat with the appropriate datum on precision parallels. This gives an accurate location and can be used to set and check the height gauge for errors.
- The angle plate ensures the component is at an accurate right angle to the surface plate.

Method of marking out

1. Place the component on the datum marked / on the parallels.
2. Set the height gauge to touch the top of the parallel, and check for variation in reading and actual size of parallels.
3. Add 20 mm to reading and scribe centreline of ∅20 hole.
4. Set height gauge 90 mm above parallel height and scribe centreline of ∅10 hole.
5. Turn component onto datum marked // set height gauge 15 mm above parallel height and scribe centreline of ∅10 hole.
6. Set height gauge 40 mm above parallel height and scribe centreline of ∅20 hole.

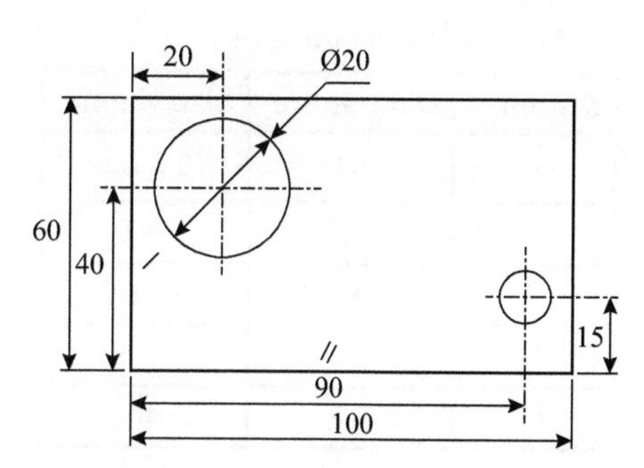

Figure 10.9 Marking out

Marking out the pitch circle diameter

The method of marking out the pitch circle diameter is to calculate the coordinates from each of the two datum faces / and //.

1. The three holes have been labelled a, b and c (see Figure 10.10).
2. The location of hole a from face / (A2) is **30** as it is on the centre line. From face // its location (C2) is the centre line plus radius of PCD = 25 + 15 = **40**
3. The locations of holes c and b require some trigonometry calculations:

- The angle cOb is 120°, the angle cxO is 90°, angle xOc is 60° and angle Ocx is 30°.
- The location of hole c from datum / is (A1) 30 – xO while the position of hole b from the same datum is 30 + side Oy, but Oy = xO
- The location of hole c from datum // is (C1) 25 – cx and the location of hole b is the same, as they have a mutual centre line.

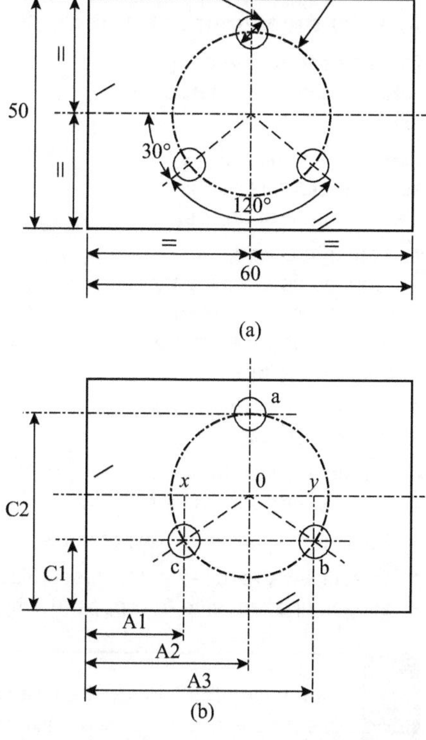

Figure 10.10 Marking out the pitch circle diameter

Calculations (see Table 10.1):

- For C1: cos = adjacent (cx) ÷ hypotenuse, cos 60° = *yb* ÷ 15 = 0.5 × 15 = 7.5, therefore *yb* = 7.5. C1 = 25 − 7.5 = **17.5**
- For A1: cos 30° = adjacent ÷ hypotenuse, cos 30° = xc ÷ 15, 0.866 × 15 = 12.99, therefore A1 = 30 − 12.99 = **17.01**, while A3 = 30 + 12.99 = **42.99**

Table 10.1 The coordinates

Datum	Coordinate	Size / mm
/	A1	17.01
/	A2	30
/	A3	42.99
//	C1	17.5
//	C2	40

- The marking out arrangements would be as shown above in Figure 10.10, using precision parallels.
- The component would be placed with datum face / on the parallels. The height gauge would be calibrated on top of the parallels and A1, A2 and A3 scribed.
- The component would be located on datum // and C1 and C2 scribed.
- Where the scribed lines intersected the centres of holes; a, b and c would be identified.
- Depending on the accuracy of location required, the holes would be machined either by drilling or boring.

Angular measurement and marking out

The sine bar is a hardened precision body with precision rollers.

Use of a sine bar allows accurate marking out and measurement of angles.

- Setting the sine bar requires determining the size of the sine bar (*l*) and the diameter of the precision rollers.
- Metric sine bars are 100, 200 or 300 mm.
- The height *h* is the value of the wrung slip gauges.

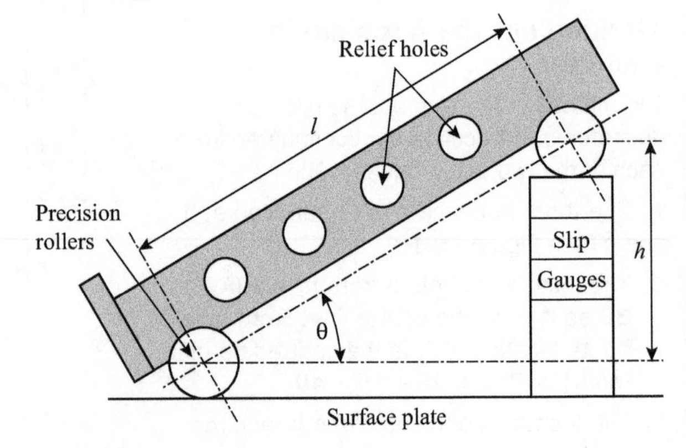

Figure 10.11 The relief holes

- So sine θ = opposite ÷ hypotenuse = *h* ÷ *l*
- To achieve a required angle θ (see Figure 10.11), and determine the wring of slip gauges needed: sine θ= opposite ÷ hypotenuse = *h* ÷ *l*, therefore *l* = *h* ÷ sine θ.

Worked example

If an angle of 30° is required with a 100 mm sine bar, *l* = 100 ÷ 0.5 = 50 mm. This means a wring of slips **50 mm** is needed.

- In Figure 10.12 the component has had a line scribed in relation to one of its datum faces.
- This arrangement can also be used to check the accuracy of an angle, but a **dial test indicator (DTI)** would be used to pass along the top of the component.

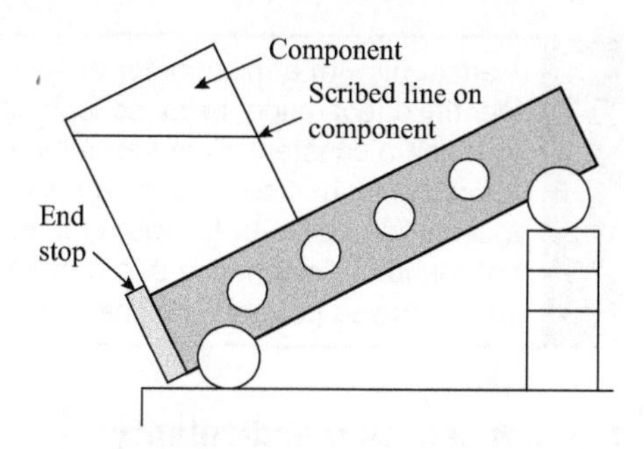

Figure 10.12 A sine bar

10.3 Inspecting work for accuracy of dimension and form

For any manufactured component there will be variations in flatness, squareness, roundness and angularity. The role of inspections is to determine if these variations are within the acceptable tolerances or not. Flatness, roundness, squareness, and concentricity relate to the **form** (shape) and **orientation** (position) and must be understood by engineers.

Flatness

Flatness can be determined by using a dial test indicator as shown in Figure 10.13. The total indicator reading is taken by passing the DTI over the length of the component, it will show if its flatness is within tolerance.

Note: Flatness will be in relation to a datum.

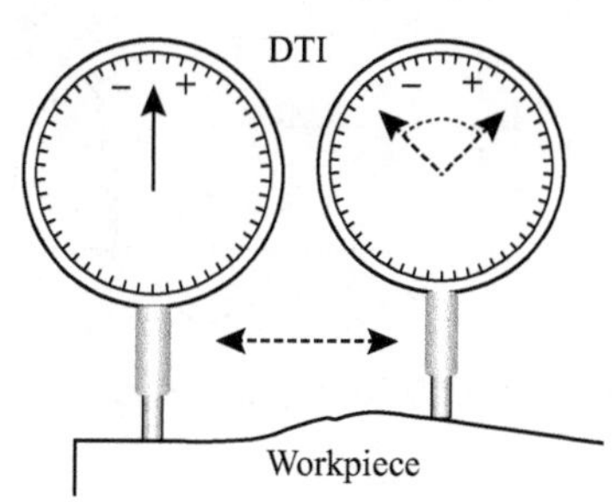

Figure 10.13 A dial test indicator

Roundness

To test for roundness (circularity) the component needs to be rotated on the axis that roundness is required.

- This is done by putting the component between centres and rotating it with the spindle of a DTI in contact with the circumference and deviations measured to see if they are within tolerance.
- A hollow component may need to be put on a mandrel for testing.
- An alternative method is to put the component in a precision V-block and rotate the component under a DTI.

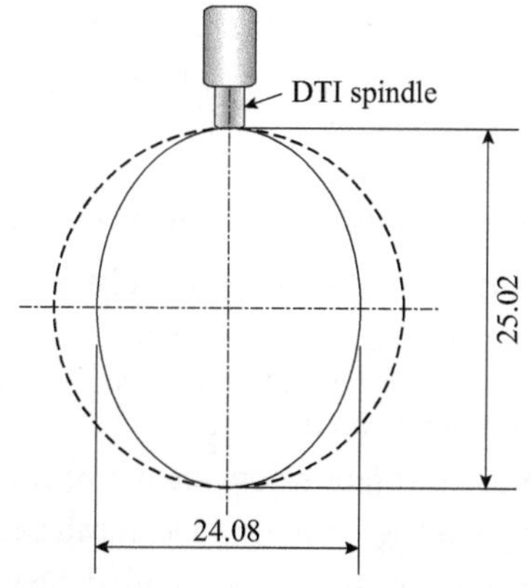

Figure 10.14 Testing for roundness

Did you know?

Measuring with a micrometer at several places around the diameter may not identify out-of-round, because lobbing during machining may produce a shape that has a constant diameter but is not round. An equilateral curved heptagon, for example, has seven curved edges and has the same diameter at any point measured across its full width (diameter). However it will be the same but it is not round. so measuring this with a micrometer would not check for roundness. These curved polygons can have a range of number of sides.

Squareness (perpendicularity)

- The squareness of Surfaces 1 and 2 (in Figure 10.15) are determined by checking with a precision square and a DTI test for perpendicularity.

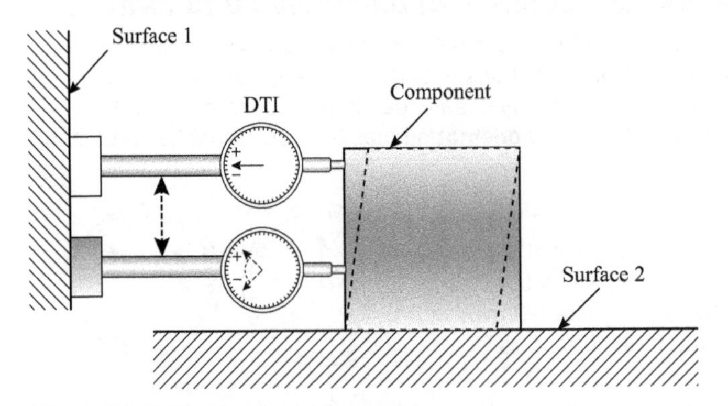

Figure 10.15 Testing for squareness

- The component is then placed on Surface 2 and the DTI is passed over Surface 1 and any deflection of the DTI pointer noted.
- It is important to ensure the component does not move during the process.
- It may be necessary to test squareness on more than one feature of the component for squareness, which may prove difficult if some faces are not perpendicular to each other.
- The dotted line indicates how the component might look if it were not square.

Angularity (being angular)

The accuracy of an angle can be checked with a sine bar as in Figure 10.12 above. Alternatively a protractor can be used and, depending on the accuracy required, a vernier protractor can be used.

- Vernier protractors are normally graduated from 0 to 90°.
- Readings can be made in two directions.
- Care must be taken to read the vernier scale in the same direction as the main scale.
- The blade can be slid round into position and clamped to the slide (see Figure 10.16).
- One side of the protractor is made flat to lie flat on a surface of the component.

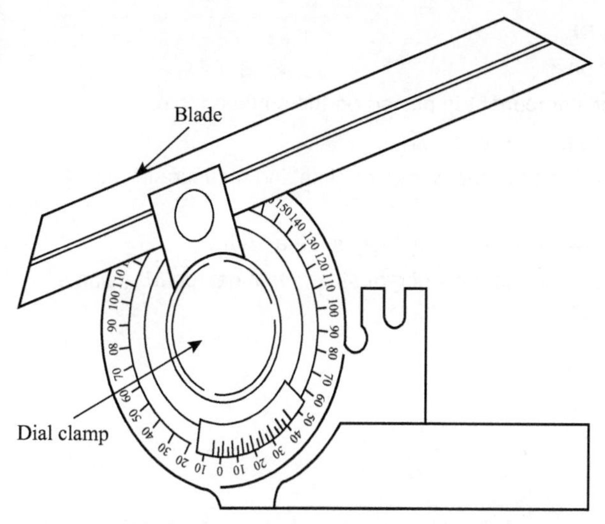

Figure 10.16 Checking angularity with a vernier protractor

Reading a vernier protractor

- Twelve divisions on the vernier scale equal 23 divisions on the main scale (see Figure 10.17).
- The 12 vernier divisions equal 23 main scale divisions; therefore, one division on the vernier scale equals $1\frac{11}{12}$ degrees.
- Because $\frac{1}{12°} = 5'$, it can be seen that $1\frac{11}{12°} = 1° \ 55'$
- The angle is read or measured by noting degrees from zero on the main scale and the zero on the vernier scale, then adding to this the number of $\frac{1}{12}$ths of a degree indicated by the coinciding division on the vernier.

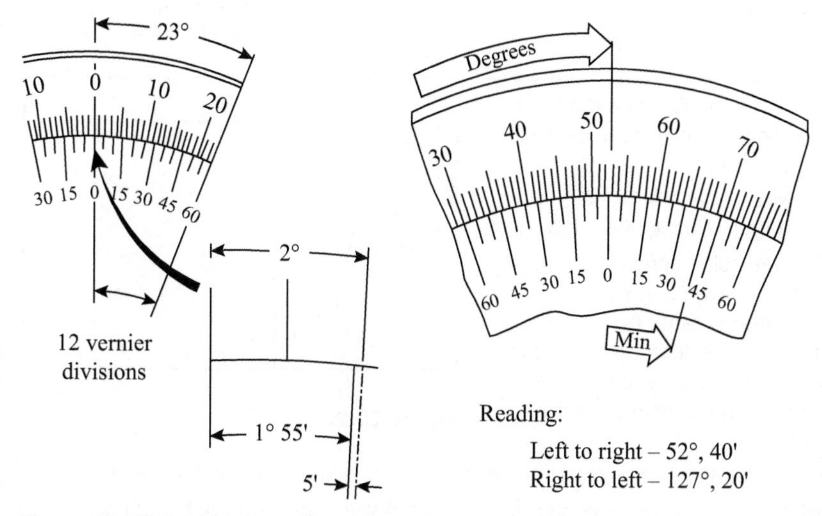

Figure 10.17 Reading a vernier protractor

Measuring an external taper

See Figure 10.18 for the dimensions:

- The smaller end of the tapered component is placed on the surface table.
- Two equal wrings of slip gauges are made to h_1.
- Place two precision cylinders on each of the wrings of slips and measure M_1 using a micrometer.
- Remove this set up and make two wrings of slip gauges equal to h_2.
- Place two precision cylinders on each of the wring of slips and measure M_2 using a micrometer.

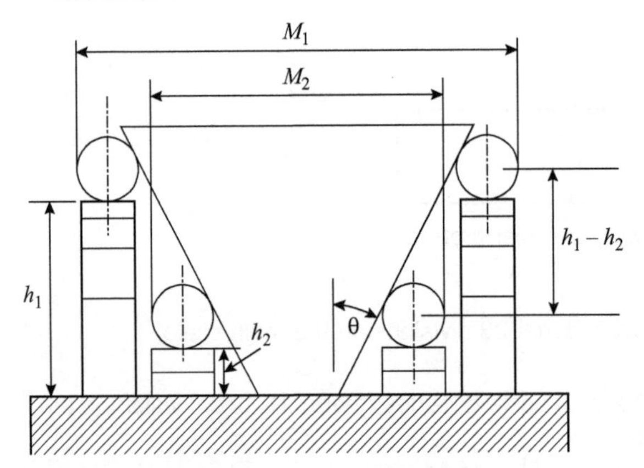

Figure 10.18 Measuring an external taper

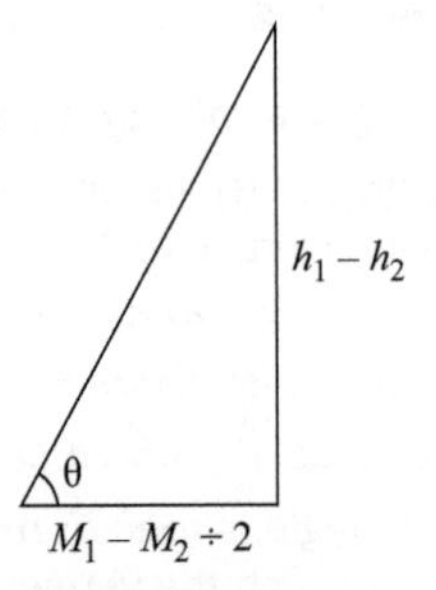

Figure 10.19 Calculating an external taper

Calculating the taper (see Figure 10.19)

- θ = half the angle of taper
- $\tan \theta = ((M_1 - M_2) \div 2)\ (h_1 - h_2)$
- Taper = $\tan \theta \times 2$

Why don't you?

1. Draw the thimble of a depth micrometer showing a reading of 19.34 mm
2. Draw a vernier protractor showing a reading of 54° 15'.
3. Two precision spheres of radius R and r have been placed in the internal taper (see Figure 10.20). The depth to the top of each sphere, h_1 and h_2, have been found by using a depth micrometer.

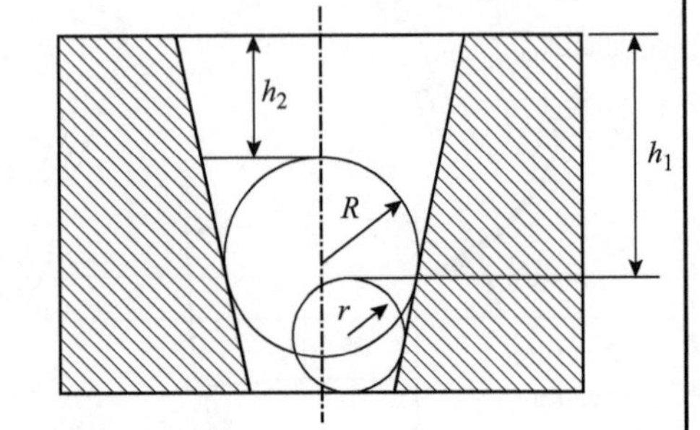

Figure 10.20

Using notation, demonstrate how the internal taper could be calculated.

Check your work with a classmate saying what went well and what could be improved. Put a copy of your work in your portfolio of evidence.

11. Operating pedestal and radial arm drill presses

Option B Section 3.6.6b

11.1 Setting machine controls

A drilling machine, sometimes called a drill press, is used to machine holes into or through metal and other materials. The cutting theory relating to the double (point) fluted twist drill is given on page 15.

This twist drill is held in the drill press by a chuck or Morse taper and rotates in a clockwise direction in plan view and is fed into the workpiece at an appropriate rate. Drilling machines may be used to perform other operations such as: countersinking, counterboring, spot-facing, reaming, and tapping. For each of these operations it is important to know how to set up the work, set speed and feed, and provide for coolant (if needed) to get the required finish.

Drilling machines are divided into two main categories:

- hand fed
- power fed.

Note: Remember drilling can also be carried out on a lathe.

Hand-fed drilling machines are the most common type of drilling machines in use and are light duty machines that are hand-fed by the operator, using a feed handle. They are either bench drilling machines or floor mounted.

Power-fed drilling machines are usually larger and heavier than the hand-fed. They are equipped with a mechanism to feed the cutting tool into the work automatically, to a preset cut speed and depth.

Cutting speed

Cutting speed refers to the number of revolutions per minute (RPM) of the drilling machine spindle, which will vary depending upon the diameter of the drill and the material being drilled (see Table 11.1).

Table 11.1 Linear cutting speed for various materials when using HSS drill

Material type	metres per min
Aluminium	75–105
Brass	45–60
Bronzes	24–45
Cast iron	18–24
Mild steel	30–38
Steel (tough)	15–18

The speed for different materials is given in surface (linear) feed per minute (SFM). Use the diameter of the drill and the cutting speed (V) to convert into (N) revolutions per minute:

$$N = \frac{1000V}{\pi D} \approx \frac{300V}{D}$$

Worked example

Find the RPM of a ∅25 mm drill cutting aluminium at a linear cutting speed of 75 m/min.

$$\text{RPM} = \frac{1000 \times 75}{\pi \times 25} \approx \frac{300 \times 75}{25} \approx 900$$

Feed rate

Feed rate is the distance the drill advances for each revolution. The feed rate used depends upon a variety of factors, including spindle power and rigidity of the machine/set up, design and type of cutting tool (reamer, etc.), and the material being cut (see Table 11.2).

An experienced operator will be able to judge the feed rate by "feel" on hand fed drilling machines.

Table 11.2 Feed rates

Material	Drilling/ mm	Reaming, etc./ mm
Aluminum	0.4–0.75	0.13–0.25
Bronze	0.4–0.65	0.07–0.25
Cast iron	0.4–0.65	0.13–0.3
Mild steel	0.25–0.5	0.07–0.25
Tool steel	0.25–0.5	0.07–0.25

Work holding

There are two main methods of **work holding:**

- table-mounted vice
- direct clamping to the machine table.

Using a table-mounted vice

As illustrated in Figure 11.1:

- Ensure the vice is seated flat on the machine table.
- Align the vice to slots in the table.
- Lightly tighten the vice to the table.

- Check vice is approximately parallel to table slots.
- Tighten bolts to secure vice to table.
- It may be necessary to seat the component on parallels if the centre of the hole needs to be perpendicular to the base of the component.

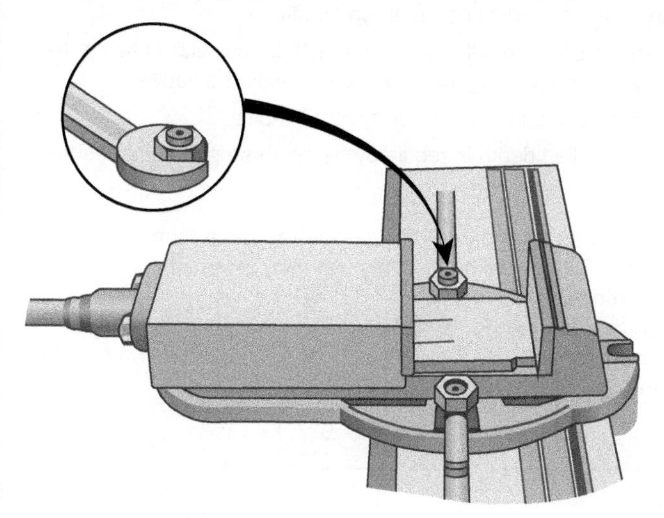

Figure 11.1 Using a vice

Direct clamping

As illustrated in Figure 11.2:

- The table slots are used to locate "T" bolts that pass through the clamp.
- One end is positioned on the component and the other on packing.
- The packing should be a similar height of the component so that the clamp is in a horizontal plane.
- Because F1 × A = F2 × B, if A is greater than B then F2 will be greater than F1, meaning the component is subject to a greater clamping force than the packing.
- Some circular components may be clamped to the table using a vee-lock.

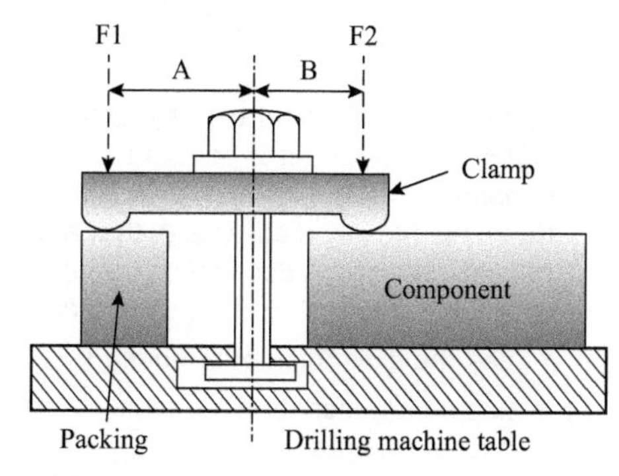

Figure 11.2 Clamping directly to the table

Starting to drill

The centre of the position of the hole is marked with a centre punch and the hole is started with a combination countersink and drill, known as the **centre drill**. The speed is calculated and the nearest available speed selected.

Other considerations may include:

- Whether a pilot hole may be needed to prevent the drill wandering during drilling.
- Twist drills usually cut oversize by about 0.02 mm. There are several reasons why drills cut oversize: rigidity of machine, wear in the spindle, leaving the drill running in the hole, and, mainly, if the drill point has been reground badly.
- Once the drill has been selected, the spindle speed needs to be calculated using the recommended linear cutting and converting this to RPM as described above.
- The best method of work holding is chosen: vice or direct clamping.
- If the hole is blind or other controlled depth is required, the depth stop will need setting.

Cutting fluids

Cutting coolants and lubricants are used in drilling to lubricate the chips formed for easier removal, aid dissipate heat generated by friction, flush away swarf, improve the finish, and to permit greater cutting speeds and feeds for greater efficiency. The fluid can be sprayed, flowed or dripped onto the component and/or tool.

See Section 2.4 cutting fluids and their uses, page 23.

Associated operations

Reaming, spot-facing, counterboring and countersinking are operations associated with drilling.

Reaming

Reaming (see Figure 11.3) is machining with a multi-edged cutting tool. It is carried out to improve the dimensional and form accuracy, and finish of drilled holes. Reaming offers the advantage that a greater number of holes can be produced with the required quality and dimensional accuracy. It requires a cutting that:

- has the exact diameter required
- has an appropriate edge profile – straight, helical-fluted and/or tapered.

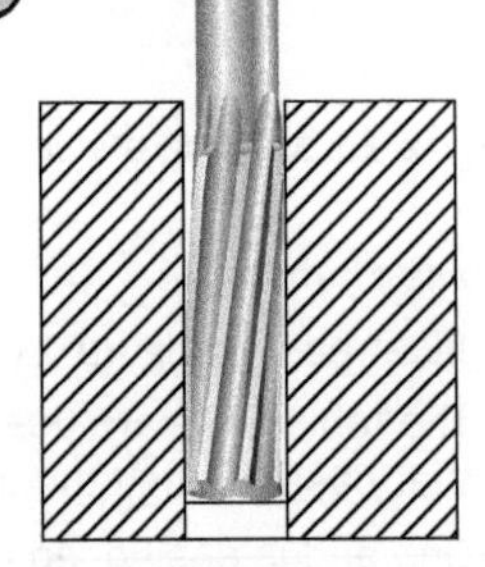

Figure 11.3 Reaming

The hole to be reamed is drilled smaller than the required finished diameter. This is called the **reaming allowance**, which is about 0.2 mm for smaller diameters up to 0.5 mm for diameters of 50 mm. A lubricant will help to produce a smoother finish to the reamed hole.

Spot-facing

Spot-facing (see Figure 11.4) is the method of smoothing and squaring the surface around a hole so as to provide a flat, square bearing surface for a nut and washer or other temporary fixing. It is generally done on rough surfaces such as castings and forgings, or angular surfaces.

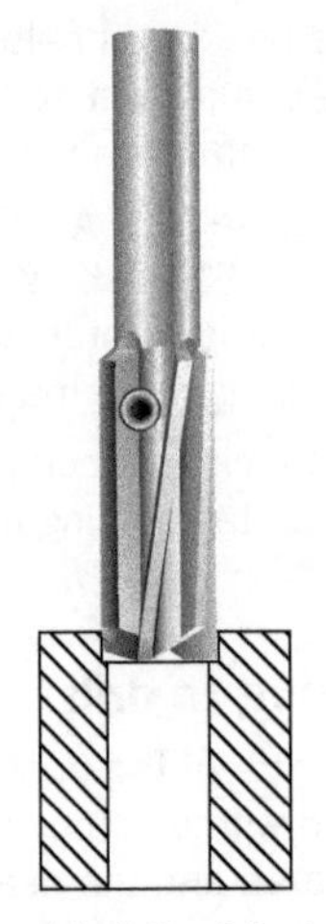

Figure 11.4 Spot-facing

The depth of spot-facing is normally about 3–5 mm, and this is usually stated on the drawing.

Drawings may state the minimum amount of metal to be left after machining.

Sometimes spot-facing is carried out with the same tool as counterboring, but to a shallower depth. The same cutting speed, feed, and lubricant can be used.

Counterboring

The counterboring tool (see Figure 11.5) has a pilot that guides the cutter into the existing hole. The hole will be pre-drilled and may have been reamed to provide an accurate location for the counterboring pilot.

The main purpose is to accommodate the head of a fastening mechanism, such as the head of a bolt or nut, so it is flush with or below the surface of the component. The depth of counterbore will be determined by the size of the fastening head. The speed will be the same for the same diameter of drill and material, and the feed rate will be between 0.02–0.05 mm per revolution.

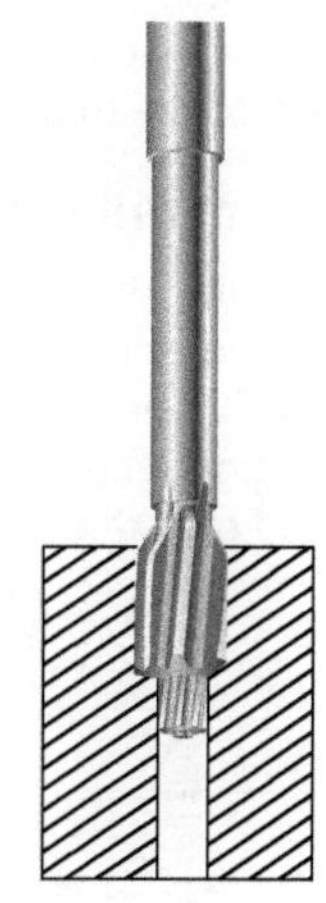

Figure 11.5 Counterboring

Countersinking

Countersinking (see Figure 11.6) uses a conical cutter and is similar to spot-facing and counterboring in that it provides a seating for the head of a screw, a countersunk screw.

The included angle of the countersink tool varies to suit the type of screw head and may be 82° or 120°. Some countersink tools have an included angle of 60°, which is the same as that of the centre on the tailstock of a lathe.

As a guide, the cutting speed should be half that of the drilling speed. The feed rate should be light, but should avoid chatter while cutting. Setting a depth stop will be necessary to ensure the screw head is flush with the component surface.

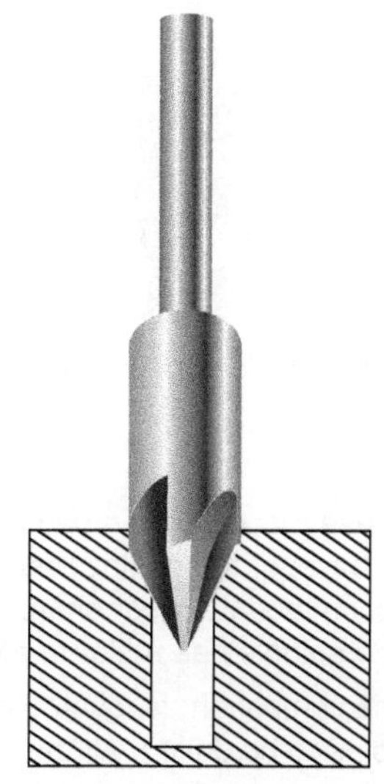

Figure 11.6 Countersinking

Machining to depth (blind hole)

The depth of a blind hole is measured to the depth of the full diameter of the drill, and excludes the included angle of the point of the drill. The depth of the hole is achieved by starting to drill until the point of the drill has just achieved its full diameter on the surface of the component. The machine is then stopped, and the required depth set on the depth stop on the machine. If more than one hole is to be machined, the depth will be checked with a depth gauge before the other holes are machined and any adjustments made. A more simple check is to insert a drill of the same size shank first and the depth read on a steel rule.

Radial arm drilling

A radial drilling machine has a geared drill head mounted on an arm that can be pivoted around its central column (360° if there is space) to the extent of its arm reach (see Figure 11.7). The drilling head of the machine can be moved along the arm, adjusted in height, and rotated. The geared drill head has power feed. The drill head can be positioned over the workpiece without the need to reposition it to drill additional holes.

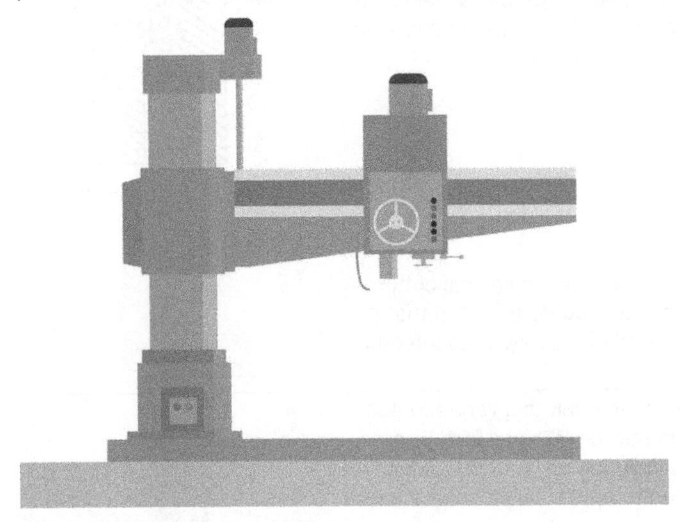

Figure 11.7 Radial arm drilling

These machines can be bench- or floor-mounted. The size of is determined by the length of travel of the drilling head along the arm. The spindle has a Morse taper that will accommodate taper shanked drills or chucks, with a tapered shank to fit into the spindle. The spindle has a slot to allow tools to be removed from the Morse taper using a drift.

The component is located on the table and the drill head is moved to the locations of the holes to be drilled. All the operations carried out on a pedestal drill press can be carried out on a radial arm drill, but to a much larger diameter of up to about Ø7 cm. Coolant and lubricant are normally available in a tank with a pump system to spray or constant flow onto the workpiece.

1. **a)** Research the main components of a radial arm drill, see Figure 11.7 above.

   b) Copy and label Figure 11.8, naming each of the main parts.

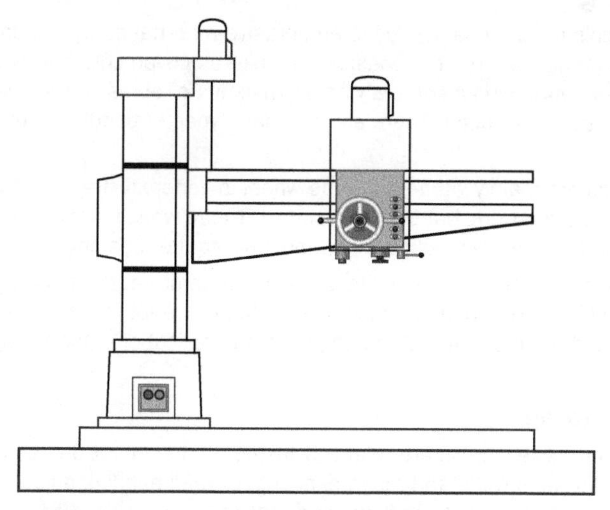

Figure 11.8 Outline of a radial arm drill

   c) Draw the arrangement for direct clamping a component with a swan neck clamp, showing how a greater force is exerted on the component rather than the packing.

   d) Calculate the cutting speed in RPM for a ∅35 mm twist drill, when machining mild steel.

   e) If that hole was then to be reamed ∅38 mm, what would be an appropriate cutting speed?

   f) Suggest how you would achieve the appropriate feed rate for reaming.

2. Figure 11.9 shows a cylindrical component that is to be drilled in the vertical plane.

   a) Draw a clamping arrangement to allow drilling to take place.

   b) If more than one hole is to be drilled on the same centreline say how you would ensure accuracy of alignment of the holes.

For both activities, work with a class mate saying what went well and areas for improvement.

Put a copy of your work in your portfolio of evidence.

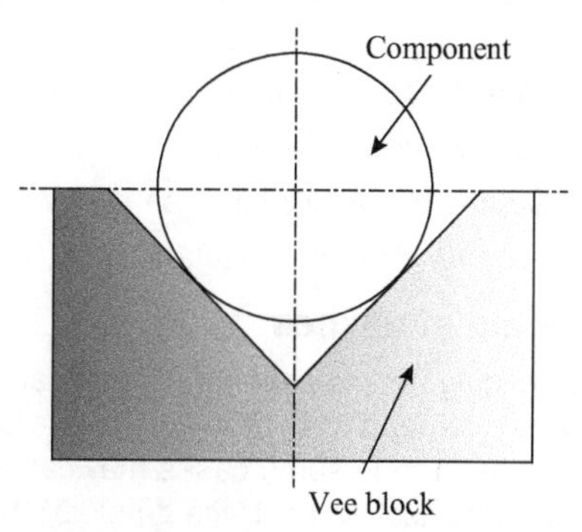

Figure 11.9 Component to be drilled

Why don't you?

12. Grinding operations

Option B Section 3.6.6c

12.1 Grinding

Grinding is a machining process that uses an abrasive material as the cutting tool, usually in the shape of a wheel. It is possible to make precision cuts and produce very smooth finishes. An **abrasive wheel** is a wheel consisting of abrasive particles bonded together with various substances. There are two main types of bonding agent: inorganic and organic.

- **Inorganic bonds** are mainly vitrified, i.e. the wheel is generally fired in a furnace to give the bond a hard, strong but brittle structure. These wheels are used for precision grinding applications, as they hold their shape, but require dressing.
- **Organic bonds** are not fired but are cured at low temperature; the bond agents are **resinoid (B), rubber (R)** and **shellac (E)**. Such wheels are tough, shock-resistant and self-dressing, and are most suited to non-precision applications, for example fettling and cutting off.

Handling and storage

- **Handling** – all abrasive wheels are relatively fragile. It should not be assumed that organic bonded wheels will stand rough handling. Careful handing guidelines should be observed to avoid chipping, cracking, and breakage.
- **Storage** – suitable racks, bins or compartmented drawers should be provided to accommodate the various types of wheel used to prevent damage so as to avoid failure during use.

Training

The Abrasive Wheels Regulations contain a training schedule that covers:

- each class of abrasive wheel to be mounted by the appointed person
- approved advisory literature
- hazards arising from use, and precautions to be observed
- storing, handling, and transporting of abrasive wheels
- inspecting and testing abrasive wheels and checking for damage
- function of all components: flanges, washers, etc., and knowledge of correct and incorrect assembly of all components
- dressing abrasive wheels
- adjusting the work rest of an abrasive wheel
- requirements of the regulations.

Grinding machines

The grinding process generates substantial amounts of heat. A **coolant** is used to prevent overheating or being out-of-tolerance. The coolant benefits the machinist as the heat generated in some cases may cause burns. In very high-precision grinding machines (most cylindrical and surface grinders), the final grinding stages are usually setup so that they remove about 0.002 mm per pass.

Types of grinding

There are many types of grinding machines, however, the four major ones are:

- offhand grinding
- surface grinding
- cylindrical grinding
- centreless grinding.

See Section 2.3 Techniques used in sharpening tools, page 20.

Features of abrasive wheels

Full details are given in Clause 5.5 of BS ISO 525: 1999.5.

Table 12.1 shows how to read the blotter on an abrasive wheel.

a) **Abrasive** means the type of abrasive used in wheel construction.

b) **Grain/grit size** means the particle size of abrasive grains. The range is expressed by number (very coarse at 4 to very fine at 1200).

c) **Grade** represents the tenacity with which the bonding material holds the abrasive grain in a wheel. Wheels are graded as 'soft' or 'hard' according to their degree of tenacity. The grade scale is expressed in letters from A (extremely soft) to Z (extremely hard).

d) **Structure** means the level of porosity in the wheel, the higher the number, the greater the level of porosity.

e) **Bond type** means the bonding material used in the wheel construction.

Table 12.1 Reading the blotter on an abrasive wheel

a	b	c	d	e	f
A	46	J	10	V	T-5
Abrasive	Grain size	Grade	Structure	Bond	Profile

Peripheral speed

As the peripheral speed of an abrasive wheel increases, the forces acting on it increase by the square of the speed. The speed at which the abrasive wheel revolves is, therefore, extremely important. It cannot be too strongly emphasised that doubling the speed increases fourfold the stress in the wheel, and hence the risk of the wheel bursting.

Did you know?

It is a requirement to report the bursting (failing) of an abrasive wheel in use as a dangerous occurrence to the health and safety authorities.

Maximum operating speed

The maximum operating speed is marked on every wheel in two ways:

- The **peripheral surface speed** which is given in m/s
- The **rotational speed** which is given in rpm. As the wheel wears down in use, the effective peripheral surface speed will reduce if the rotational speed remains constant and may result in a reduced grinding efficiency. To counteract this, the spindle speed can be increased, providing the maximum peripheral surface speed of the wheel is not exceeded. It is essential that the spindle speed is reduced to its original value before fitting new wheels.

Inspection

Ring test

- Suspend the wheel from the hole on a finger. For larger, heavier wheels, rest them on a pin or other way that will allow it to "ring" if struck.
- Tap the side of the wheel gently with a non-metallic tool midway between its edge of centre.
- Listen for the sound (ring) that comes from the wheel when it is tapped. Turn the wheel 90° and repeat the test.
- Any wheel with a dull sound must be set aside and marked as "do not use".

Balancing abrasive wheels

- Abrasive wheels are balanced by the manufacturers, so new wheels should not need balancing.
- Wheels that have been frequently trued may become out-of-balance and the forces associated too excessive.
- One technique is to use heavy paint for slight out-of-balance, but this is a very specialised process.
- The most frequently used method is to fit flanges with adjustable weights that can be positioned at any point within an arc.
- The wheel is fitted onto a mandrel (see Figure 12.1) that sits on two knife points and allowed to come to rest – the heaviest point will fall to the bottom.
- The weights are moved to the top to counteract this heavy side, and the wheel is turned and allowed to come to rest – again the heaviest point will fall to the bottom.
- This process is repeated until the wheel is at rest at any point.

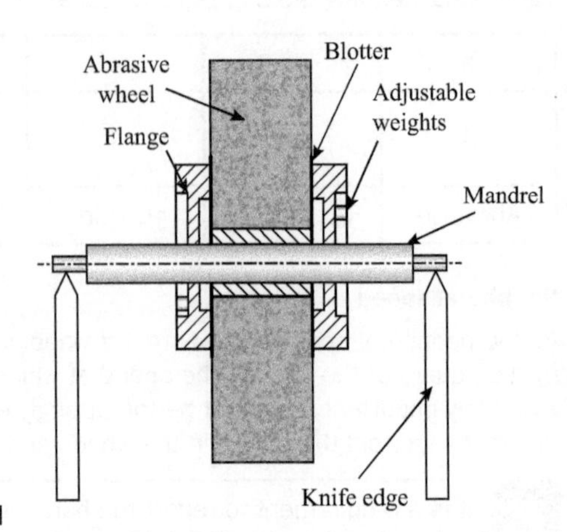

Figure 12.1 Balancing abrasive wheels

- The wheel is then mounted back on the machine with the adjustable flanges in place. Note: Blotters are placed between the wheel and its flanges to absorb and diffuse any clamping forces.

Truing abrasive wheels

The primary purpose for truing of abrasive wheels is carried out to return the wheel to its required shape or geometry, so that accurate grinding can be achieved whether a profile or flat surface is required. The secondary purpose may be to achieve concentricity of the wheel. Truing can sometimes return an out-of-balance wheel back into balance.

Truing is normally carried out by using an industrial diamond which can be done on the machine, or on a machine specifically for truing wheels.

Dressing abrasive wheels

Dressing is carried out to recondition the wheel's cutting surface and is used to:

- expose the grit from the bond and so improve cutting
- splinter the grit and make them sharper and more free cutting
- remove build up of residue of previously ground material.

The most commonly used method involves using a star wheel dresser which is positioned on the tool rest, ensuring the two lugs are on the tool rest, then moved across the abrasive wheel producing a square surface.

12.2 Attaching and aligning workpieces for grinding

Surface grinding

- Surface grinding (Figure 12.2) produces flat, parallel, and angular surfaces, by feeding work in a horizontal plane beneath a rotating abrasive wheel.
- Ferrous workpieces are held magnetically to the table, and are mostly ground by traversing the table.
- Work is aligned by using a dial test indicator (dti) plunger to detect any deflection of the pointer as the workpiece is traversed.

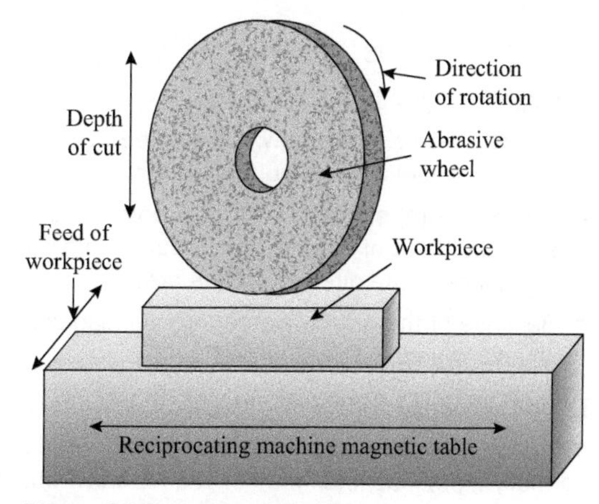

Figure 12.2 Surface grinding

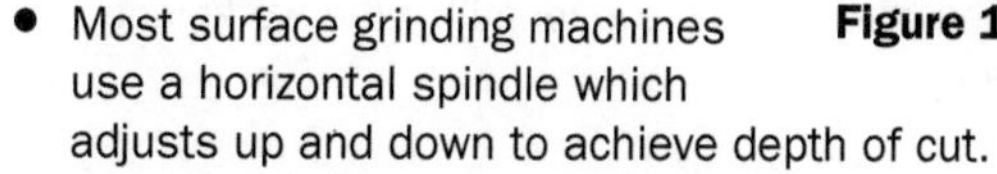

- Most surface grinding machines use a horizontal spindle which adjusts up and down to achieve depth of cut.
- The periphery, or side, of the abrasive wheel may be used to remove material.
- Both the periphery and side of the wheel can be used to produce an accurate surface finish to a shoulder.
- Some surface grinding machines have an abrasive wheel that rotates in a horizontal plane (vertical spindle) with a recess in the wheel so that flay surfaces can be ground.

Cylindrical grinding

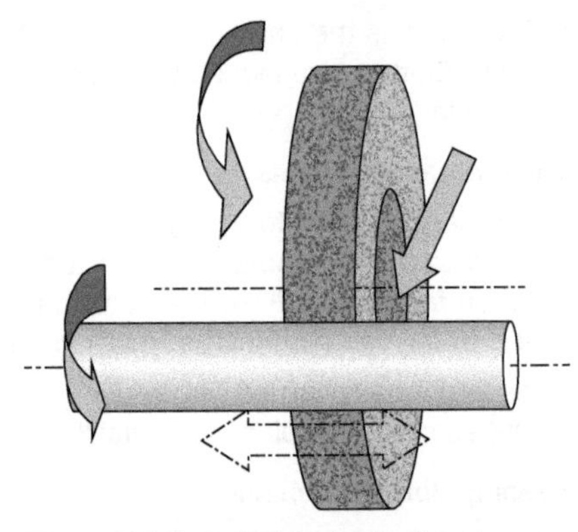

Figure 12.3 Cylindrical grinding

- In Figure 12.3, the workpiece rotates about the centre (fixed axis) and the surfaces machined are concentric to that axis of rotation.
- Cylindrical grinding produces a cylindrical surface that may be either parallel or tapered.
- The taper is usually generated by an offset tailstock method.
- The cylindrical grinding machine includes an abrasive wheel head, which incorporates a cross-slide, that moves the wheel head to and from the workpiece.
- A headstock drives the workpiece, and a tailstock locates and supports the traversed workpiece.
- At the headstock, a chuck or catch plate can be used.

Centreless grinding

Centreless grinding is similar to centred grinding except that there is no spindle. There are three main types of centreless grinding:

- through-feed grinding
- in-feed grinding
- end-feed grinding.

Through-feed grinding

- Figure 12.4(a) shows how, through-feed grinding, the workpiece rotates between the grinding wheel and a regulating wheel.
- Figure 12.4(b) shows how one, or both wheels, of the centreless grinding machine are canted out of the horizontal plane. This gives a horizontal feed direction to the workpiece in the direction of the arrow.
- Because of the axial traverse through-feed centreless grinding, it can only have right circular cylindrical workpieces.

In-feed grinding

In-feed grinding is quite different to through-feed grinding in that the workpiece is normally a pre-machined profile. The abrasive wheel is dressed to the shape required, and the width of the abrasive wheel must be greater than the length of the workpiece. This process cannot produce right cylinders.

End-feed grinding

During end-feed grinding (see Figure 12.5), the workpiece moves axially between the grinding and regulating wheels up to an end stop and then moves out again. The wheel can be dressed to form quite complex profiles.

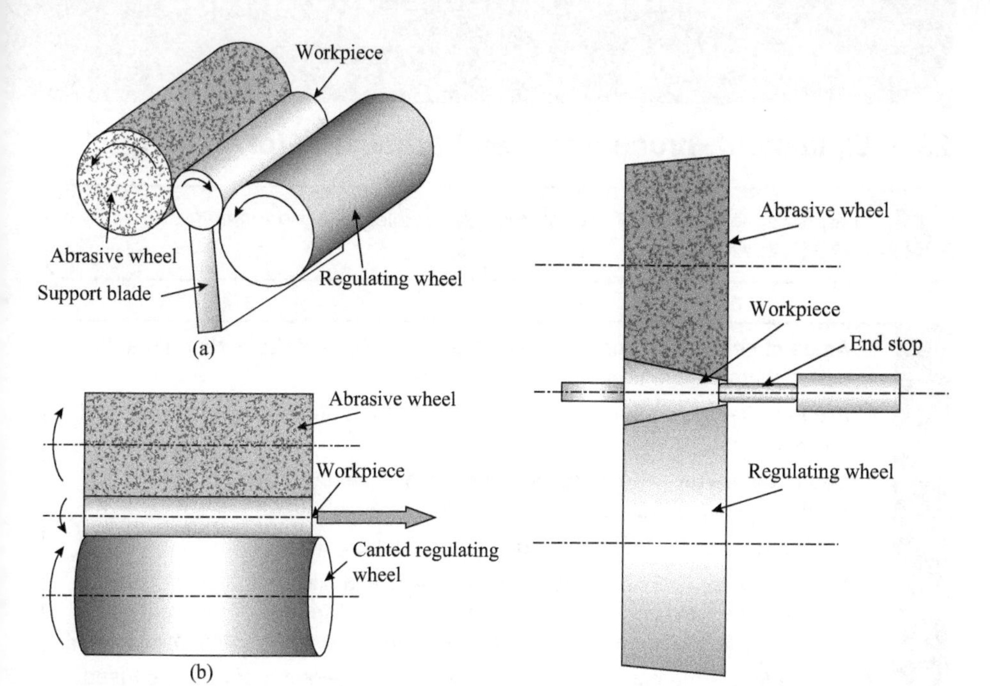

Figure 12.4 Through-feed grinding

Figure 12.5 End-feed grinding

Why don't you?

1. Table 12.2 gives information from an abrasive wheel blotter:

Table 12.2 An abrasive wheel blotter

Abrasive	Grain size	Grade	Structure	Bond	Profile
RA	60	K	7	V	T-5

   a) Explain each piece of information, e.g. RA = ruby aluminium oxide, what does each of the others mean?
   b) Research what grinding situations this type of abrasive wheel might be used for.

2. Magnetic chucks on a surface grinder can only hold ferrous metals. Describe and draw an arrangement for surface grinding a brass workpiece.
3. Explain the difference between truing and dressing, giving an example of each application.

13. Performing welding operations

Option B Section 3.6.4

13.1 Equipment, procedures, and standards for welding

The precise definitions of the following terminology can be found in ISO 857-2:2005.

Key terms

- **Angle of bevel** – the angle to which the end or edge of the parent metal is cut or chamfered in preparation for welding.
- **Arc length** – the distance between the end of the arc welding electrode and the weld pool.
- **Deposited metal** – the metal produced by the melting of the filler rod or electrode which forms the resultant weld.
- **Filler rod** – rod of metal that is deposited during welding.
- **Flux (welding)** is a combination of carbonate and silicate materials used to shield the weld pool from atmospheric gases. The heat of the weld pool makes the flux melt and excludes atmospheric gases preventing oxidation.
- **Fusion face** – the part of the edge or face of the parent metal that is fused with the filler rod during welding.
- **Fusion zone** – the place where the parent metal and deposit metal (filler rod) have fused and formed a weld pool.
- **Parent metal** – the material of the component that is to be welded.
- **Penetration** – the depth of the fusion zone.
- **Run** – the length of the weld pool for one pass of the filler rod or electrode.
- **Spatter** – globules of molten metal thrown out during welding.
- **Weld pool** – the pool of liquid metal formed during fusion of parent and deposit metals.
- **Welding sequence (direction)** – the order and direction in which welds and runs are made.

Types of welding

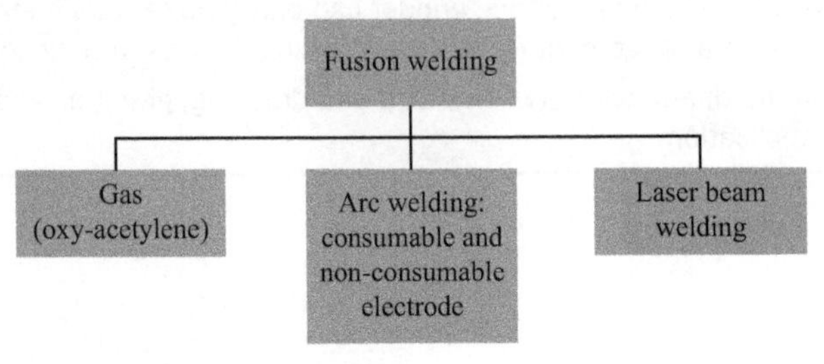

Figure 13.1 Types of welding

Fusion welding is a generic term for joining together materials, and normally (but not always) involves the melting (fusing) of a filler rod or a consumable electrode with a parent metal by heating to high temperatures to form a molten weld pool.

The three main types of fusion welding are: oxy-acetylene, arc welding, and laser welding (see Figure 13.1).

Oxy-acetylene welding

Oxy-acetylene welding uses combustion of acetylene and oxygen with a flame temperature of about 2700 °C. This can be used to weld steel with a thickness up to about 5 mm. A filler rod similar and compatible with the parent metal is used to make the join. The process can be used for welding aluminium and thinner gauge steel, and requires a high level of skill.

See **Table 1.3** Common fuel and oxidiser pairings, page 10.

Gas welding equipment

- Gas cylinders (see Figure 13.2) are colour-coded: oxygen is black and acetylene is maroon.
- Cylinders should be used upright and fastened to prevent accidental knocking over.

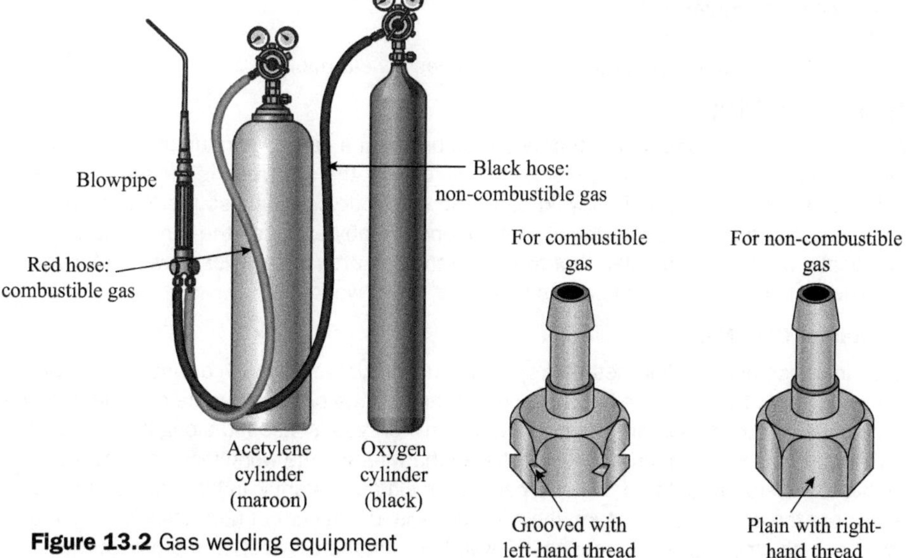

Figure 13.2 Gas welding equipment

Figure 13.3 Threaded connectors

- The oxygen and acetylene hose carry the gases and are also colour-coded.
- The hoses connect to a torch with a range of nozzle sizes marked with their gas consumption.
- The correct size of nozzle depends on the material type and thickness.
- The threaded connectors (Figure 13.3) follow the coded theme by having different-handed threads, and plain and grooved.

- Goggles must have the correct protection for welding.
- The appropriate clothing must be worn, including gloves.

Arc welding

A range of welding processes are based on heating with a simple electric arc welding, also known as **stick welding**. The process uses an electrical source, normally a.c., through a holder which carries an electrode coated with a flux. An earth cable connects the workpiece to the welding machine to provide a return path for the current (see Figure 13.4). The weld is initiated by "striking" the tip of the electrode against the workpiece which initiates an electric arc. A temperature of 6000 °C, generated almost instantly, produces a weld pool.

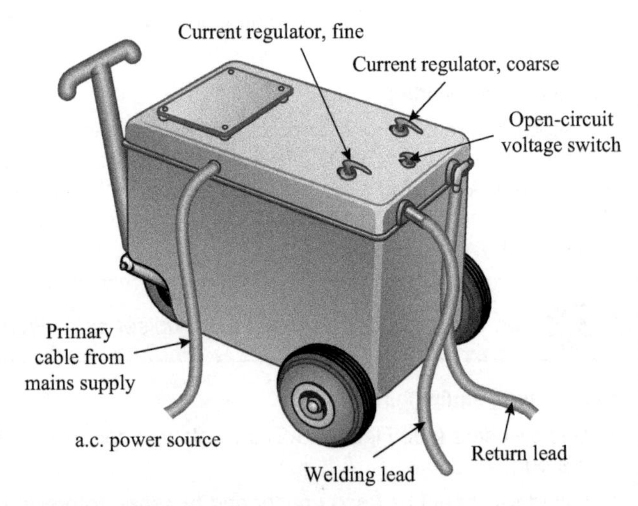

Figure 13.4 Arc welding equipment

Manual arc welding

Manual arc welding uses an arc that is struck between a coated consumable electrode and the workpiece (parent metal). The metallic core of the electrode is melted by the arc and is transferred into a molten weld pool. The electrode coating also melts and forms a gas shield around the arc and the weld pool, and thereby excludes the atmosphere from the cooling weld pool. The slag that forms is removed after each layer of weld. Manual arc welding is low cost, flexible and suitable for repair work.

Metal Inert Gas (MIG)

Metal inert gas uses a filler metal (wire) on a spool and driven through a tube into the arc. The large amount of filler wire on the spool means that the process can be considered to be continuous with long, uninterrupted weld runs. An inert gas is also fed along the tube and exits around the wire. An arc is struck between the wire and the workpiece and because of the high temperature of the arc a weld pool forms almost instantly. In this process the key issues are the correct gas mixture, the filler wire speed and current (amperage). Once these have been set, the skill level required is lower than with the oxy-acetylene process.

Tungsten Inert Gas (TIG)

Tungsten inert gas uses a non-consumable tungsten electrode with an arc struck between this and the workpiece surface. An inert gas is used to shield the weld area and filler rod used. The process is suited to joining non-ferrous metals, including aluminium. The process requires a high degree of skill, but high quality welds can be produced.

The power source can be either alternating current (a.c.) or direct current (d.c.) – normally a.c. is used. When the source is from the mains supply, the voltage (in volts, V) must be reduced to between 40V and 100V. However, the mains source amperage needs to be increased to give the required output for welding.

There are three main output ranges:

- 150–200A
- 250–300A
- 400–500A

The equipment must be set up by a competent person, ensuring it is adequately earthed. The a.c. output is taken from the second side of the transformer through a current regulator (see Figure 13.5):

1. Connect welding lead to electrode holder.
2. Connect welding to output terminal of power source.
3. Connect return lead to earth clamp and workpiece.
4. Connect return lead to return terminal of power source.

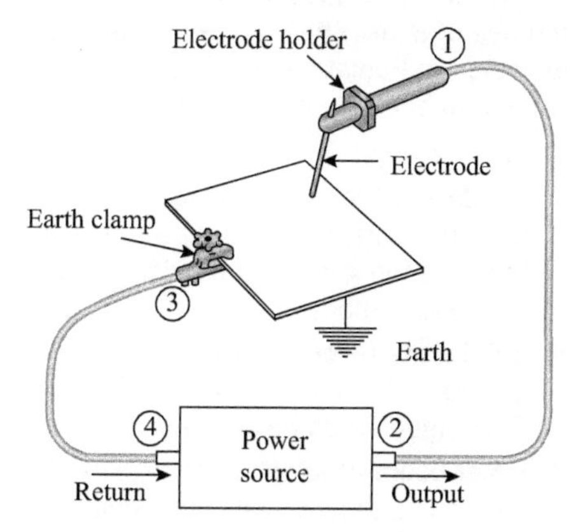

Figure 13.5 Cable connections

Welding visors and helmets

The light emitted by arc welding includes invisible ultraviolet and infrared rays, so screening of operators and others in the vicinity is needed for their protection.

- A visor, helmet or screen is essential for personal protection. These can have an adjustable headpiece (as in Figure 13.6) and have a hand shield.
  Note: Goggles alone are not adequate protection for arc welding.
- The screen must be of an approved filter.
- The visor and helmet should give protection to the eyes, face, and throat. A hand shield will also giving protection to the wrist and hand.
- The specifications for screen filters are covered by: ISO 4852:1978 Personal eye-protectors – Infrared filters and ISO 4851:1979 Personal eye-protectors – Ultraviolet filters.
- Clothing must be free of oil, grease and other flammable material.

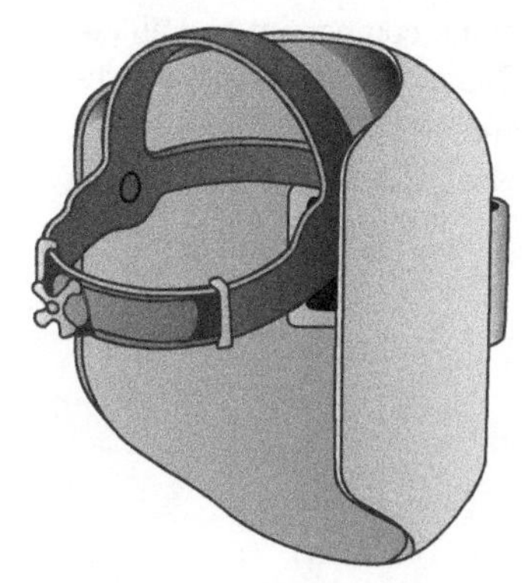

Figure 13.6 A welding helmet

- Protective clothing should give cover from throat to the knees with a suitable leather apron.
- Protection includes gloves that cover the hands and wrists and suitable footwear.
- Fume extraction and ventilation are important considerations for all types of welding, but particularly electric welding. Positive air pressure masks or mechanical extraction or ventilation are a must in confined spaces such as welding booths.

Laser welding

Laser welding is a technique where materials, usually metal, are joined together through the use of a laser beam. **Laser** is the abbreviation of Light Amplification by Stimulated Emission of Radiation, and is a non-contact process that requires access to the weld zone from one side of the parts being welded.

The principle types of laser used in welding are:

- **Gas lasers** that use a mixture of gases such as helium and nitrogen. There are also carbon dioxide lasers. These lasers use a low-current, high-voltage power source to excite the gas mixture using a lasing medium and operated in either a pulsed or continuous mode.
- **Solid state lasers** operate at one micrometre wavelengths. They also can be either pulsed or continuous mode. Pulsed operation produce joints similar to spot or stitch welds (see notes below), but with complete penetration.
- **Laser beam welding** is used for materials that are difficult to weld using other methods and for small components. Inert gas shielding is needed for more reactive materials. Helium and argon are fully inert gases which do not react with the weld metal. By contrast, other welding gases or its components, such as nitrogen, oxygen and carbon dioxide, are reactive and may affect mechanical properties of the weld.

Notes:

Stitch welding, sometimes called intermittent welding, is not a continuous weld along a joint, but a weld that is broken up by spaces or gaps between welds, which results in a "stitch"-like appearance.

Spot welding is used to join two overlapping pieces of metal at small points or spots by application of pressure and electric current. An electrode is positioned on each side of the component and an electrical current passed through while pressure is applied. This causes the metals to fuse together locally in a spot. This process is repeated at appropriate intervals.

See Table 13.1 for a summary of the three types of welding.

Table 13.1 Welding advantages and disadvantages

Type of welding		Advantages	Disadvantages
Oxy-acetylene		• Easy to learn. • Lower costs compared to other types of welding. • More portable than arc welding. • Oxy-acetylene equipment can be used for flame cutting.	• Welds appear rougher in appearance than other methods – finishing may be required. • Produces larger heat zones which may affect mechanical properties adversely. • Safety issues around oxy-acetylene relating to naked flames and storing gases.
Arc welding	Manual	• Versatile, suitable for a wide variety of welding positions and applications. • Low level skills requirement.	• High levels of fume produced, extraction may be required. • Selection of consumables and settings may result in lower weld strength.
	MIG	• Minimum of welding fume. • Consistently high quality welds with minimum operator skills. • Suited for thick sections welds. • Suitable for welding carbon, low alloy and alloy steels.	• Welds limited to flat or horizontal position. • Possible problems maintaining correct electrode alignment.
	TIG	• Suitable for a very wide range of materials. • Especially good for welding thin sections and small workpieces. • Produces high quality welds of good appearance.	• Generally restricted to flat or horizontal welding. • High level of skill required.
Laser welding		• Works well with high alloy metals. • Small heat-affected zone. • Dis-similar metals can be welded. • No filler metals are necessary. • Produces deep and narrow welds. • Small, thin components can be weld. • No contact with material.	• High capital cost for equipment and maintenance costs. • Rapid cooling rate may cause cracking in some metals.

13.2 Welding joints and techniques

These are some of the types of welding: tack, flat, horizontal, vertical, overhead, leftward and rightward, fillet.

When the arc is struck (Figure 13.7), the voltage falls to the "arc voltage". This will be between 20 and 25 V for typical electrodes at normal arc length – equal to the diameter of the core wire.

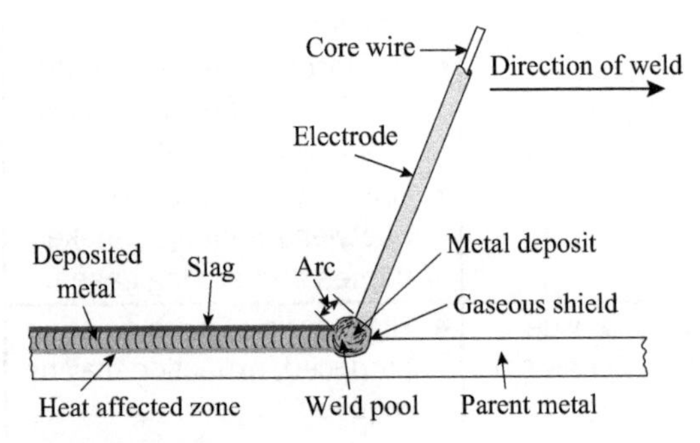

Figure 13.7 The arc gap

The arc voltage depends upon type of electrode use and arc length.

When the arc is struck, the core wire and the immediate area of the workpiece quickly reach fusion temperature. As the core wire melts and molten metal transferred across the arc gap and fuses with the workpiece.

Tack welding

Tack welds are used to position the components to be welded accurately prior to welding, so that they will be held in the correct location in relation to each other during welding. The tacks are either consumed by the final weld or removed by grinding and/or chiselling.

Flat welding

Flat welding is the preferred term; sometimes called **down-hand**. The arc is stuck at the left-hand edge with the electrode at 90° front to back and 65° to 75° in the direction of weld. (See Figures 13.8 and 13.9.)

Note: The centre line of the length of the weld is called the **axis** of the weld.

- If L is about $1\frac{1}{2}D$, there is a larger weld pool, poor penetration, the weld bead is course and flat with larger splatter.
- If L is about $\frac{1}{2}D$, there is intermittent arc with, uneven weld bead, slag on spatter globules.

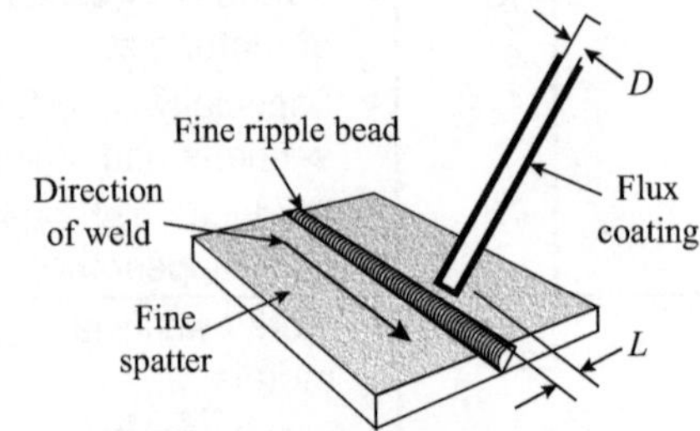

Figure 13.8 Flat welding

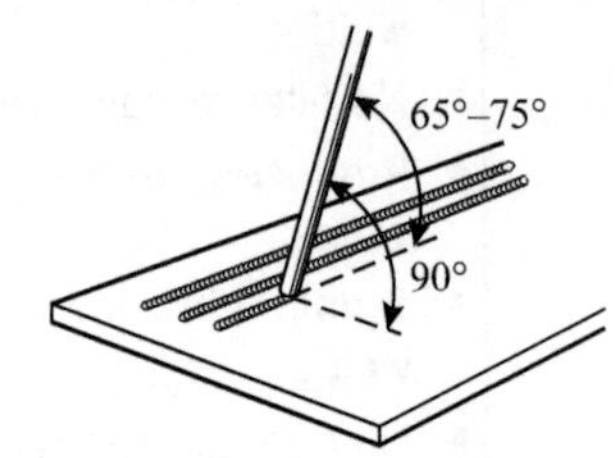

Figure 13.9 Angles in flat welding

Horizontal welding

- In horizontal welding, the weld axis is approximately horizontal.
- The nozzle of the torch should be at an angle of 60°–70° and directed to heat and fusion of both edges of the joint.
- The white tip of the flame's cone should be about 3 mm above the weld pool.

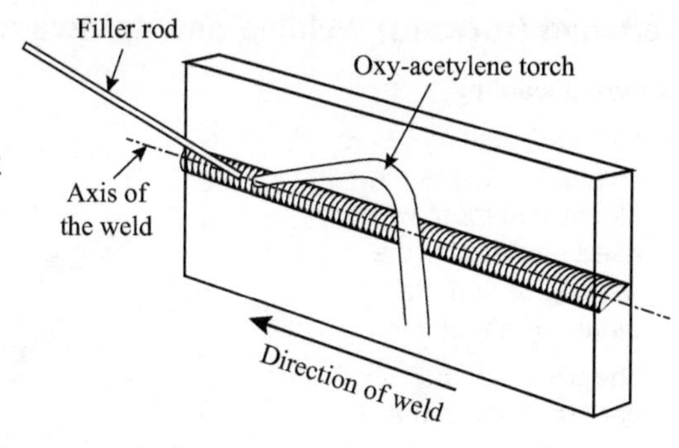

Figure 13.10 Horizontal welding

Vertical welding

- Welding on a vertical surface is more difficult than welding in the flat position, because gravity forces the molten metal to flow downward.
- For arc welding in the vertical position the current settings should be less than those used for the same electrode in the flat position. The current settings for welding upward (Figure 13.11(a)) on a vertical surface are a little higher than those used for welding downward (Figure 13.11(b)) on the same surface.

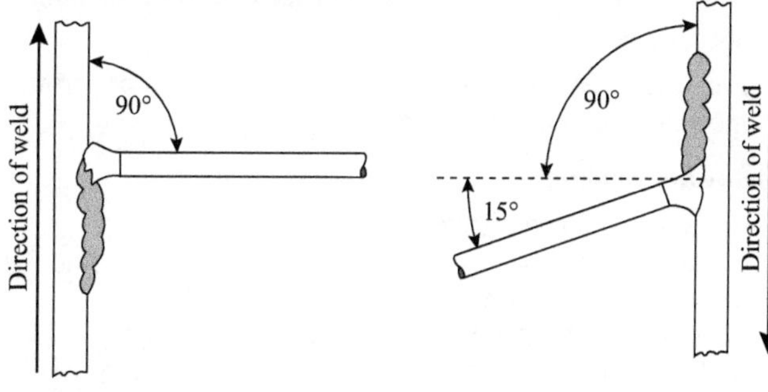

Figure 13.11 Vertical welding

- The appropriate angle between the electrode and the parent metal should be as shown, in order to deposit a good bead weld when welding vertically.

Overhead welding

- This is the most difficult welding position because gravity makes the molten pool flow downwards even more than in vertical welding.
- It requires a very short arc to be maintained to ensure control of the molten pool.
- When bead welding, the electrode should be held at an angle at 90° to the parent metal as shown in Figure 13.12.

The electrode can be inclined up to 15° in the direction of weld, as shown, to provide a better view of the arc and molten pool.

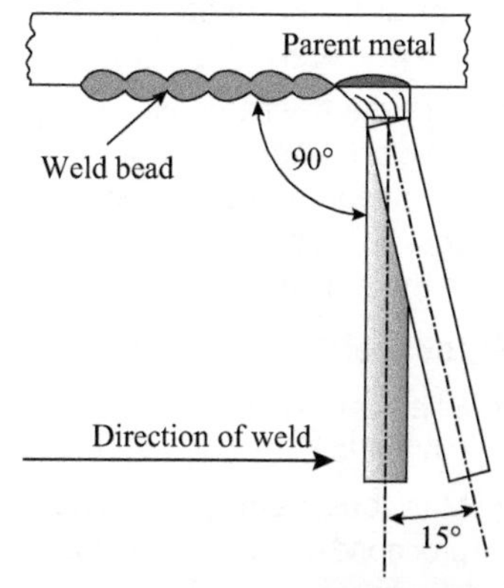

Figure 13.12 Overhead welding

Leftward (forward) welding and rightward (backward) welding

Leftward welding

- Leftward welding (Figure 13.13) is the oldest and most widely used method for the welding of thin steel plates of about 4 mm.
- The filler rod is generally about the same thickness of the parent metal.
- It is an economic process with good mechanical properties if the parent metal is not too thick.

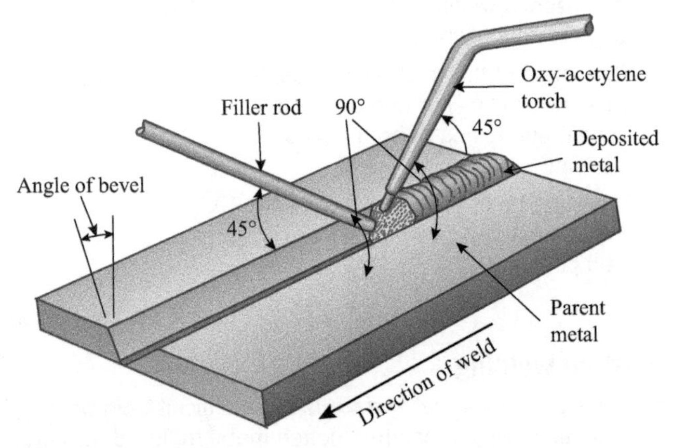

Figure 13.13 Leftward welding

- Both the torch and filler rod should be at angles of 45° from left to right and 90° to the work surface front to back.

Rightward welding

- Rightward welding (Figure 13.14) consists of commencing at the left-hand side of the plate and proceeding towards the right, the filler rod following the torch.
- The torch and filler rod are held at similar angles to those used in leftward welding for similar thicknesses of parent metal.
- Because the flame is directed towards the "V" ahead of the molten pool no side-wise motion of the welding torch is necessary, meaning a narrower groove can be used.

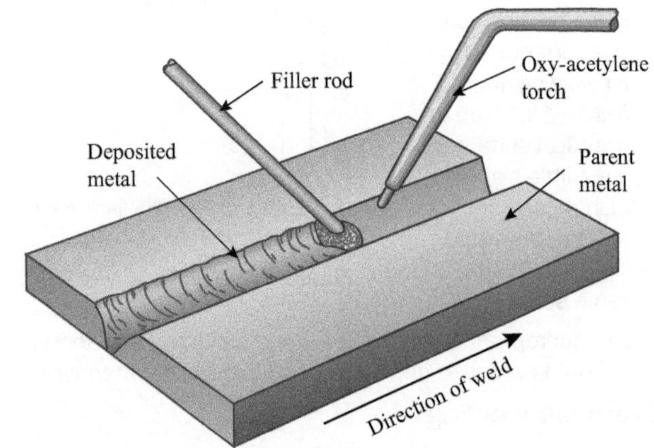

Figure 13.14 Rightward welding

Fillet weld

- Fillet welded joints, such as 'T', lap and corner joints, are the most common joints in fabrication.
- Many other joining techniques use some form of a fillet joint including non-fusion processes such as brazing and soldering.
- In Figure 13.15, L = length of leg and A = throat thickness.

Welding joints and symbols

The standards for welding symbols are covered by ISO 2553 and BS 499 Part 2.

The arrow and reference line indicate the location of the weld, see Figure 13.16.

Additional information can be added where necessary, such as length of leg, throat, etc. The length of a fillet weld leg is usually shown before the welding symbol. Unless otherwise stated, the leg length is the same as the plate thickness where plates are of the same thickness.

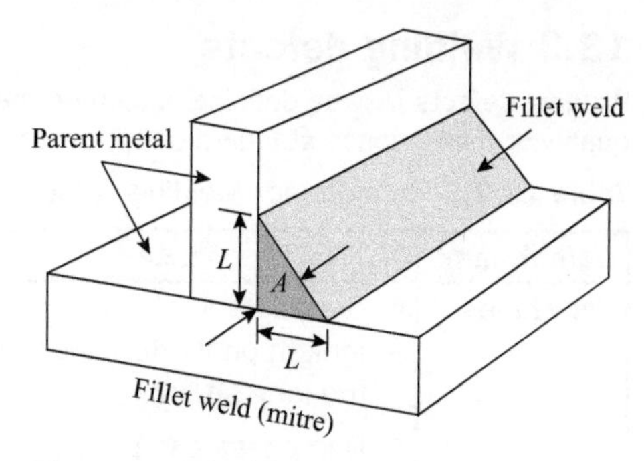

Figure 13.15 A fillet weld

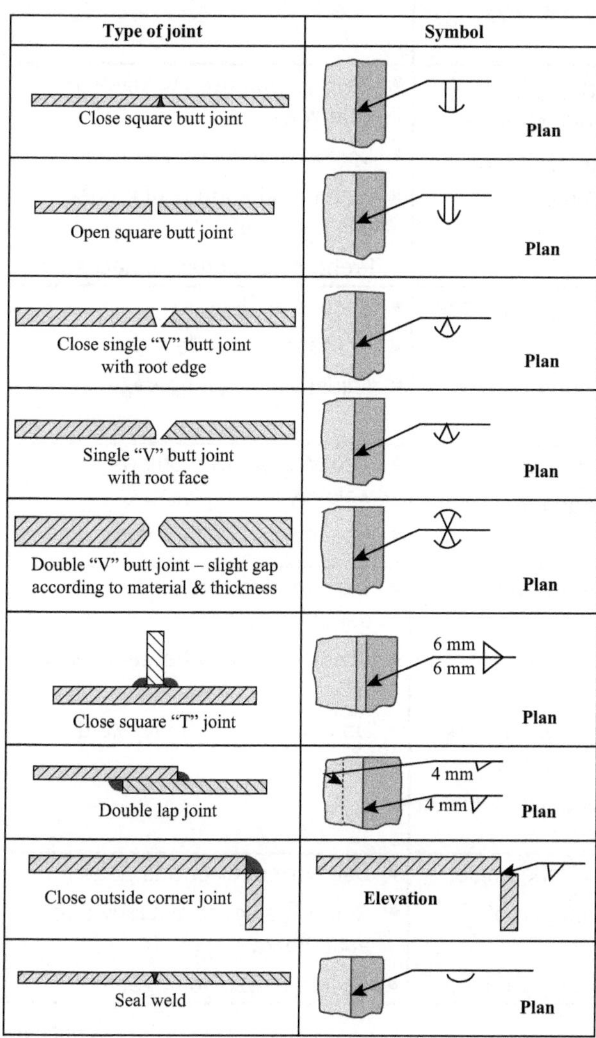

Figure 13.16 Symbols for different types of joint

13.3 Welding defects

Welding defects may be defining outcomes that failed to meet the minimum acceptance quality or performance standard or specifications (see Table 13.2).

Table 13.2 Some common welding defects

Defect name	Cause	Remedy
Blow holes	• Gases trapped in the molten pool often caused by too long an arc. • Dampness can be a contributing factor. • Shielding gas not covering molten pool.	• Reduce length of arc. • Remove moisture from welding area. • Ensure shielding gas covers molten pool.
Cold cracking	• Rapid cooling. • Wrong electrode type. • Grease, paint, or dirt on the surface of weld.	• Reduce /remove hydrogen in heat zone. • Correct coating on electrode. • Clean weld surface of grease, paint, rust, etc. • Preheat component if necessary.
Excessive spatter	• Too high welding amperage. • Long arc length. • An acute electrode angle.	• Reduce arc length. • Increase electrode angle. • Adjust welding amperage.
Hot cracking	• Impurities during welding. • Too much carbon in filler rod.	• Use MIG or TIG methods of welding. • Use low carbon filler rod.
Inclusions	• Often associated with lack of penetration. • Also associated with flux shielding. • Trapped solid matter: slag, oxide, foreign matter. • Ineffective cleaning between beads/runs.	• Discard electrodes with damaged coating. • Clean effectively between weld runs. • Do not lay down heavy weld beads.
Lack of fusion	• Incorrect electrode angle. • Slag contamination of molten pool. • Incorrect joint preparation.	• Use correct electrode angle. • Remove slag, particularly after each bead. • Ensure correct joint geometry.

(*Continued*)

Table 13.2 Some common welding defects (*Continued*)

Lack of penetration	• Incorrectly prepared weld joints. • Fast traverse speed. • Too large electrode diameter. • Too low an arc current.	• Ensure correct joint geometry. • Select correct electrode diameter. • Use correct welding amperage.
Undercut	• Too high a welding temperature melting parent metal away. • Too short an arc length.	• Reduce welding temperature. • Reduce arc length.

Destructive (DT) and non-destructive testing (NDT)

Destructive testing includes methods where the material is stressed until it fails, or is cut in order to determine properties that cannot be determined by visual inspection. This is a means of finding out about the quality of a weld and if it meets the required specification.

See macro- and micro-examination, page 6.

Non-destructive testing is the most cost effective testing and is visual examination both during and after welding. The finished component can be sent to the customer if it passes inspection.

Methods include:

- **Magnetic particle testing**, where a magnetic field is used to locate surface and near-surface welding imperfections.
- **Dye penetrants** that are solvent- or water-based liquids used to locate surface-breaking defects in all non-porous metals.
- **Radiography** passing X-rays through the component onto a film that captures internal images including any imperfections.

Why don't you?

1. While electric welding normally uses a.c. equipment, describe the basic equipment required for d.c. rectifier output welding equipment.
2. Figure 13.17 shows six welding symbols. Describe each one and produce a drawing showing their application.

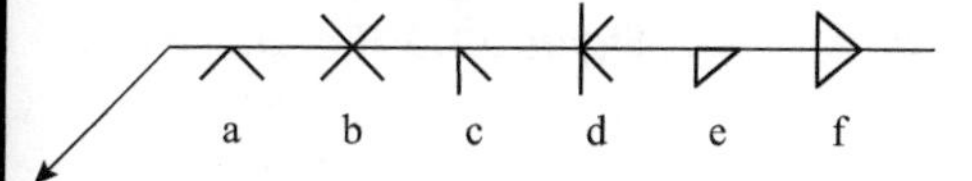

Figure 13.17 Welding symbols

14. Milling machine

Option B Section 3.6.6e

14.1 Setting up horizontal milling machines

Horizontal milling machines have a horizontal spindle with a cutter mounted in the same plane on an arbour above a right to left (X–Y) table. A universal horizontal milling machine has a table that features a vertical rotary function for machining slots or grooves at different angles. The design of the horizontal milling machine allows robust and rapid removal of material off the workpiece.

Horizontal milling cutter selection

See page 16 for the features and nomenclature of a side and face milling cutter.

- **Side and face mills** have teeth that cut both on the periphery and both sides. They are used for cutting slots.
- **Staggered tooth mills** also cut on the periphery and side but the side cut is on alternate teeth. The staggered-tooth design prevents chips from interfering with the cutting action.

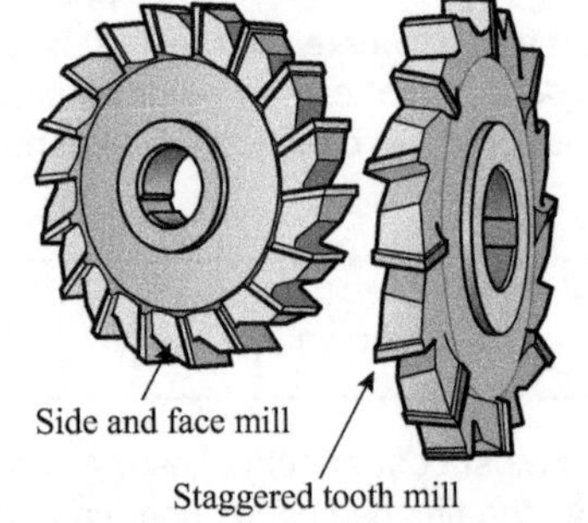

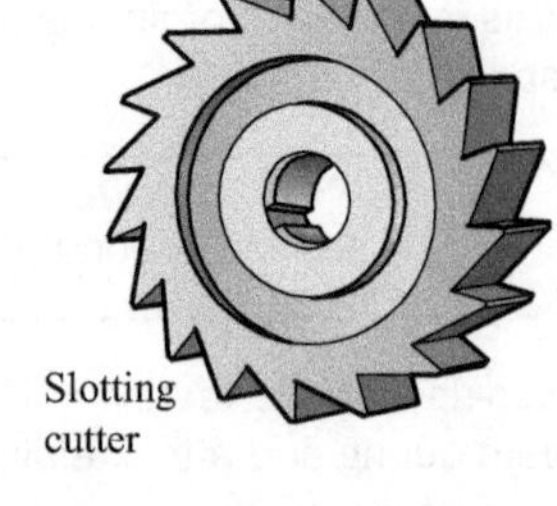

Figure 14.1 Milling cutters

- The **half-side milling cutter** cuts on the periphery and one side. It is used to machine up to a shoulder.
- The **slotting cutter** cuts only on the periphery. They are used for milling slots, keyways and narrow flat surfaces.
- The slitting saw (Figure 14.2) is used to cut material to length, mill narrow slots, and undercut shoulders.

Figure 14.2 The slitting saw

General cutter selection

- Use the smallest diameter suitable.
- Use the most suitable profile of cutter.
- Consider the material to be cut.

Work holding

Setting up a machine vice is shown in Figure 14.3:

- If there is a universal table ensure it is set to zero.
- Position machine vice approximately parallel with the table.
- Lightly clamp the vice to the table, tightening one bolt more to act as a fulcrum and allow slight movement of the vice.
- A precision parallel is gripped in the vice jaws.
- A dti with its magnetic base is placed on the vertical slide of the machine.
- The plunger of the dti is positioned to run without fouling on the vice.
- The machine table is traversed longitudinally and any deflection of the dti pointer noted.
- The vice is moved until no deflection of the dti pointer is noted.
- The vice is tightened in place and placed perpendicular to the machine vertical slide, checked and adjusted if needed.
- The same procedure is followed if the vice is to be located across the machine table.

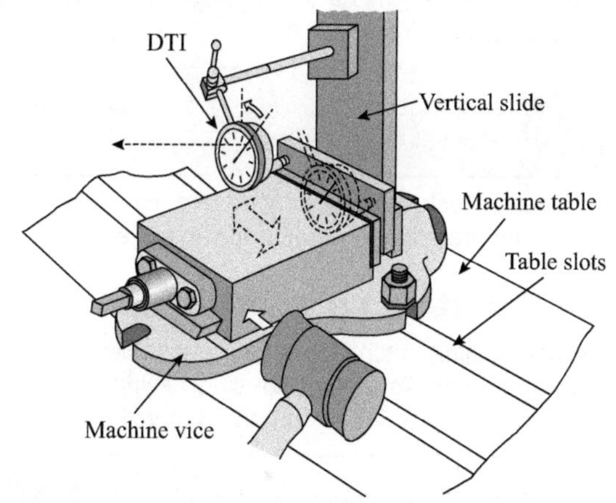

Figure 14.3 Setting up a machine vice

Direct clamping to the table

How to clamp to a table:

- Ensure the universal table is set to zero.
- Wring together slips equal to the width of the table slots.
- Put clamp "T" bolts in table slots.
- Position wrings of slips into table slots.
- Position precision parallel against the wrings of slip.
- Test for parallelism with the table using a dti and traversing the table (see Figure 14.4).

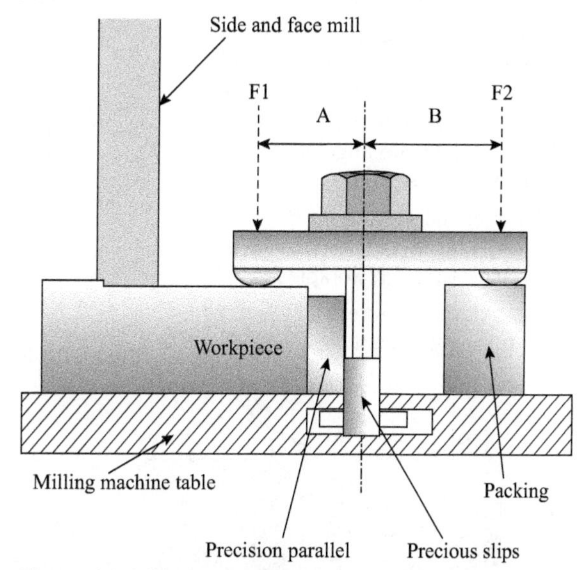

Figure 14.4 Direct clamping to the table

- A flat, square face of the workpiece is positioned against the precision parallel.
- The workpiece is then clamped as shown, ensuring distance A is smaller than B. This ensures a greater clamping force on the workpiece because: $A \times F1 = B \times F2$
- The side and face mill is positioned on the machine arbour.

Note: The closer the cutter is positioned to the nose of the arbour spindle, the greater the rigidity of the set up.

See Section 2.2, page 17 for speed in RPM, feed rate and depth of cut (metal removal rate, MRR).

- Round stock can be clamped directly using vee blocks and suitable clamps. The vee blocks are aligned parallel to the table using slip gauges and precision parallel.

14.2 Setting up vertical milling machines

Vertical milling machines have the spindle mounted vertically perpendicular to the machine table. As well as possibly having a universal table, some vertical milling machines have a universal head that will swivel in a left to right plane.

The features and nomenclature of horizontal milling cutters covered on page 114 are valid for vertical milling cutters also.

Vertical milling cutter selection

- The end mill (Figure 14.5) cuts on both the end and periphery. They are used for milling shoulders and flat edges and faces. They can be used to machine angular faces with a universal head.
- The face mill (Figure 14.6) is used to machine flat surfaces and roughing steps in workpieces.
- The slot drill (Figure 14.7) is used for milling slots and keyways. They

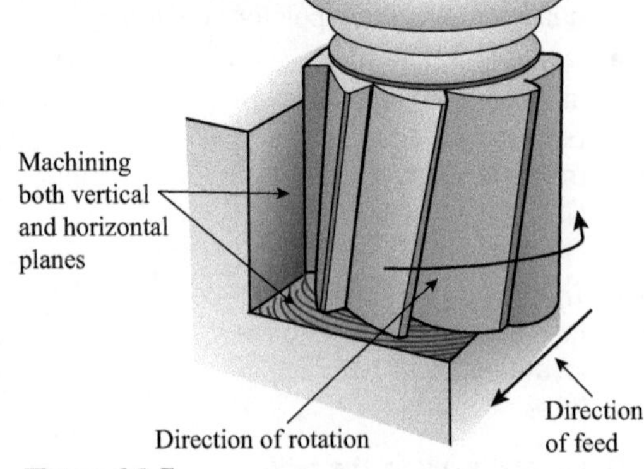

Figure 14.5

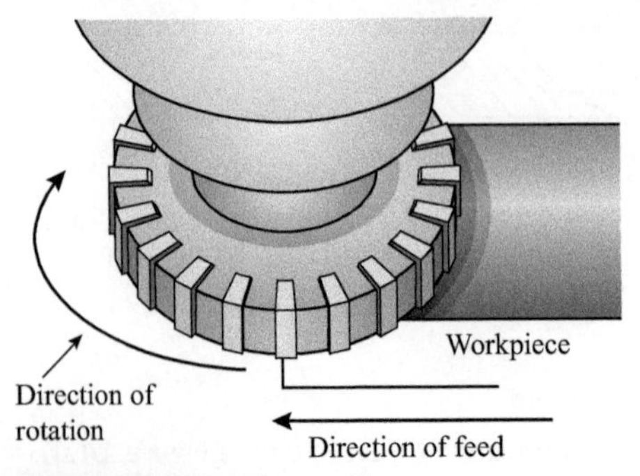

Figure 14.6 The face mill

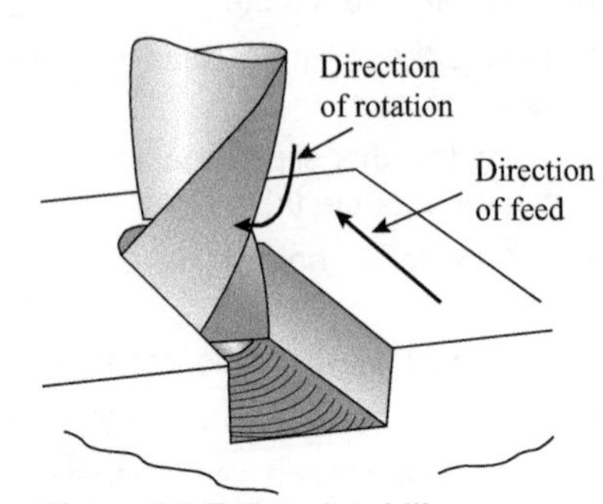

Figure 14.7 The slot drill

can be used to machine seating on angular or curved surfaces, and repositioning incorrectly positioned holes.

- The T-slot cutter (Figure 14.8) is used to cut T-slots after an end mill has been used to machine a slot to accommodate the shank of the T-slot cutter.

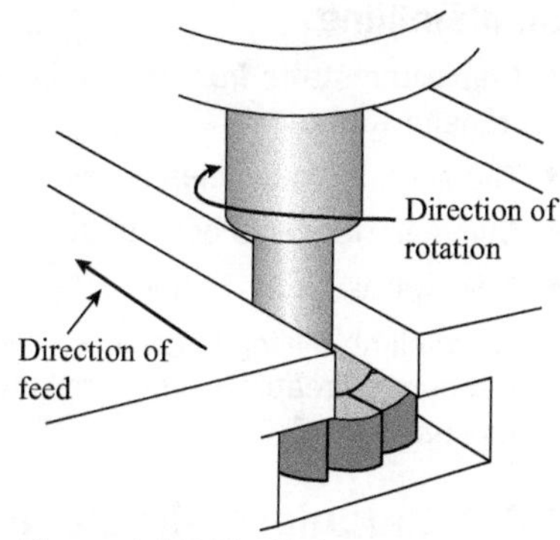

Figure 14.8 The T-slot cutter

General cutter selection

- Select a cutter the size of the slot, groove or profile required.
- If the slot or groove is not a standard width, use the largest cutter below the width of slot and remove equal amounts on each side of the slot to achieve required width.
- The choice of straight or helical teeth cutters is determined by the material being cut and the rigidity of the workpiece and/or machine.
- Helical teeth are desirable for heavy cuts.
- Straight toothed cuts are best for producing thin walls and shallow slots.

14.3 Characteristics of up-cutting and climb milling operations

There are two distinct ways to cut materials when milling, whether horizontal or vertical milling:

- Conventional (up-cut) milling as shown in Figure 14.9
- Climb (down-cut) milling as shown in Figure 14.10.

The difference between these two techniques is the relationship between the rotation of the cutter and the direction of feed. Put simply, when up-cut milling the cutter tends to push the workpiece *away* from the direction of feed, while for down-cut milling the cutter tends to pull the workpiece *in* the direction of feed.

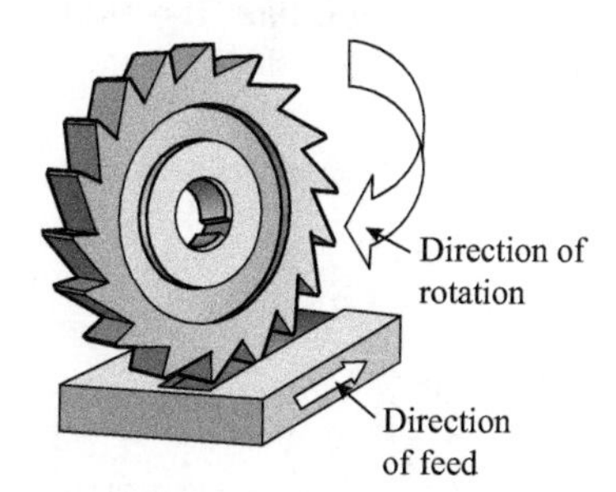

Figure 14.9 Conventional milling

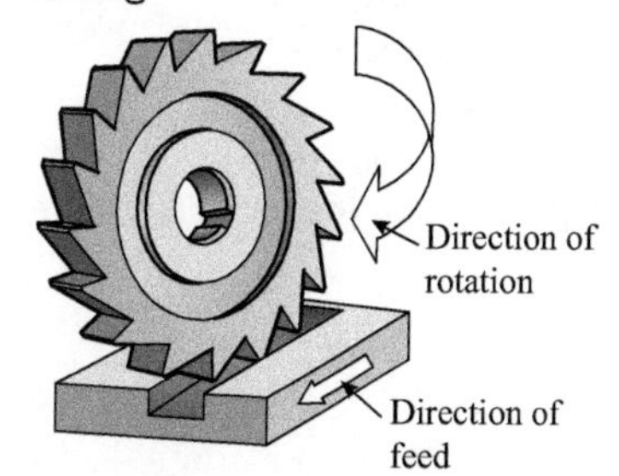

Figure 14.10 Climb milling

Conventional milling

- The chip width starts from zero and increases, which causes more heat into the workpiece which could cause work hardening.
- The tool rubs more at the beginning of the cut, causing faster tool wear and decreases tool life.
- Chips are carried upward by the tooth and fall in front of the cutter creating a poorer finish.
- Upwards forces created in horizontal milling tend to lift the workpiece.

Climb milling

- Chip width starts from maximum and decreases, so heat generated is more likely to transfer to the chip.
- The shear plane is cleaner, which causes the tool to rub less and increases tool life.
- Chips are removed behind the cutter which reduces the chance of re-cutting.
- Downwards forces in horizontal milling are created that help hold the workpiece down.
- While climb milling is favoured because it increases tool life and gives a better finish, it requires greater machine rigidity and no, or very little, backlash within the table feed mechanism.

14.4 Calculate simple indexing

The **indexing fixture** (Figure 14.11) consists of an indexing (dividing) head and tailstock (sometimes called footstock), plus a workpiece support for longer workpieces. The dividing head and tailstock are located in the T-slots of the milling machine table and held by T-slot bolts. An index plate containing a series of holes (graduations) is used to input the rotation of the index head spindle, which accommodates either a chuck or catch plate on its screwed spindle nose. The indexing handle is connected to the index head spindle through to a worm and worm wheel. Workpieces are held in a chuck or between centres by the indexing head spindle and footstock. There are many variations of the indexing fixture. The dividing head is used to produce flats, grooves or splines accurately on the circumference of the workpiece.

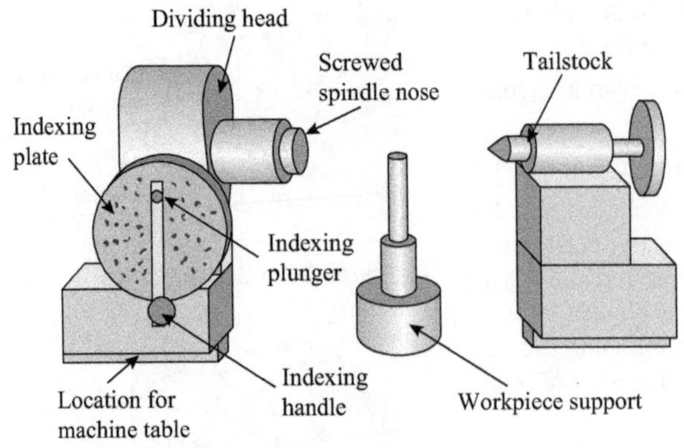

Figure 14.11 The indexing fixture

Indexing

- The worm wheel normally has 40 teeth, and one revolution of the indexing handle will rotate the indexing spindle one-fortieth of a revolution which equals 9^0 of angular displacement.
- Forty turns of the indexing handle will rotate the spindle nose one complete revolution. The worm and worm wheel ratio is 40 : 1.
- The indexing plate is inscribed with concentric circles of holes. Each dividing fixture has interchangeable indexing plates to allow a wide range of indexing.
- The indexing plunger is used to ensure the indexing handle is rotated the correct amount of rotation for the required division of the workpiece.

Calculating simple indexing

Calculating the indexing (number of turns of the indexing handle) is achieved by dividing 40 by the number of splines or flats required.

$$\text{Indexing} = \frac{40}{N} \text{ where } N \text{ is the number of divisions}$$

e.g. to machine a square on piece of round stock the calculation is:

$$\text{Indexing} = \frac{40}{4} = 10$$

This means 10 complete turns of the indexing handle are needed to machine four flats to produce a square. The indexing handle will be returned to the same hole on the indexing plate to ensure accuracy of indexing.

If a pentagon or five slots were to be machined using a dividing head, the calculation would be:

$$\text{Indexing} = \frac{40}{5} = 8$$

which means eight full turns of the indexing handle after each machining.

Assuming 16 divisions are required the indexing calculation would be:

$$\text{Indexing} = \frac{40}{16} = 2.5$$

which is two and a half turns of the indexing handle.

This is achieved by selecting a circle of holes on the indexing plate that is divisible by 2. So, if a circle of holes with 20 holes is used, then the indexing handle would be turned two full turns: the 10 holes in the 20 circle of holes on the indexing plate, equalling two and half turns of the indexing handle.

1. Figure 14.12 shows a side and face cutter climb milling, redraw the set-up showing up-cut milling.

What specific considerations do you think are necessary if climb milling is used?

The side and face mill is making a roughing cut while milling mild steel with a ⌀150 mm, 12 tooth HSS cutter.

Calculate the metal removal rate.

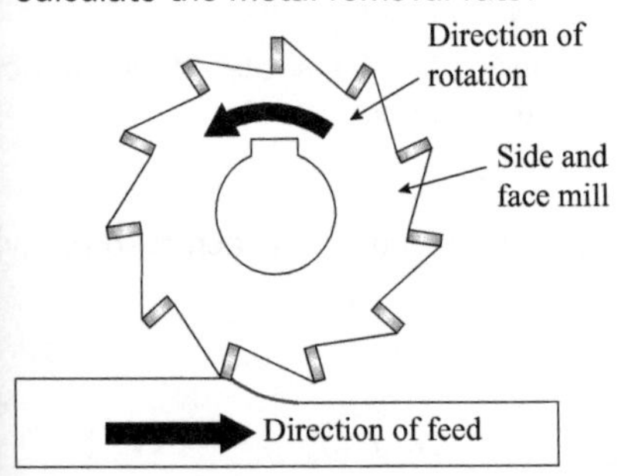

Figure 14.12

2. Figure 14.13 shows a cylindrical workpiece in a vee block that is to have a slot machined on its centreline.

Draw the set up, show and explain how the vee block will be made parallel to the machine table.

Also show the clamping arrangements and how a greater clamping force will be exerted on the workpiece.

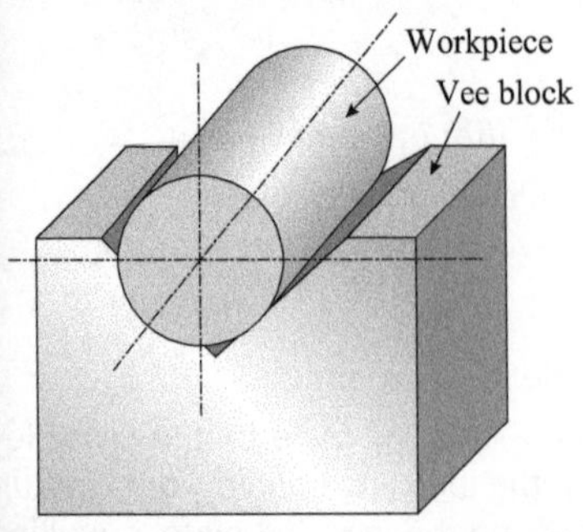

Figure 14.13

3. Calculate the indexing for machining an octagon or eight splines on a round shaft using a dividing head. Show your working. Suggest how you would set the dividing fixture up to ensure the cutter would be on the centreline of the workpiece.

Compare your work with that of a classmate. Write what went well and what could have been done more effectively. Put a copy of your work in your portfolio of evidence.

Why don't you?

Glossary

Allotropy Many metals can exist in more than one crystalline form. Above 906°C iron changes from a body-centred cubic (BCC) structure (ferrite) to a face-centred cubic structure (austenite).

Angle of bevel The angle to which the end or edge of the parent metal is cut or chamfered in preparation for welding.

Arc length The distance between the end of the arc welding electrode and the weld pool.

Austenite This is a face-centred cubic (FCC) unit cell and is only possible in carbon steel at high temperature. It has a structure which can contain up to 2% carbon in chemical combination.

Carbon An element that strengthens steel and gives it the ability to be hardened by heat treatment. Carbon forms compounds with other elements called carbides, e.g. iron carbide, chrome carbide etc. It has a small atom size which fits into clusters of iron atoms.

Cementite A very hard material, but when mixed with soft ferrite layers its average hardness is reduced considerably.

Deposited metal The metal produced by the melting of the filler rod or electrode which forms the resultant weld.

Driven The gear on a shaft that is turned by the driver (sometimes called the follower).

Driver The gear on the shaft that supplies power to the gear train.

Eutectoid point The lowest possible temperature of solidification for any mixture of specific constituents. In terms of an alloy of iron and carbon, it is the lowest melting/solidification point of any other alloy composed of the same constituents in different proportions.

Ferrite This has a body-centred cubic (BCC) structure which can hold very little carbon, typically 0.0001% at room temperature.

Filler rod A rod of metal that is deposited during welding.

Flux (welding) This is a combination of carbonate and silicate materials used to shield the weld pool from atmospheric gases. The heat of the weld pool makes the flux melt and excludes atmospheric gases preventing oxidation.

Functional dimensions These are the sizes, shape and limits and tolerances permissible of features of a component that will affect the component's performance in use.

Fusion face The part of the edge or face of the parent metal that is fused with the filler rod during welding.

Fusion zone The place where the parent metal and deposit metal (filler rod) have fused and formed a weld pool.

Idler gear A gear positioned in between the driver and driven gears (sometimes called an **intermittent gear**).

Inspection The engineering process by which the size, shape and function of a component are checked for accuracy and form (and safe performance where appropriate), to ensure the component has been manufactured within specification.

Knife tool The standard knife tool is the most commonly use single point cutting tool.

Limits and fits These are a set of rules and conventions relating to the dimensions and tolerances of mating machined parts to achieve the required class of fit.

Metal removal rate This is the measurement for how much **metal** is removed from a part in relation to time.

Output gear The name for the last gear in a train.

Parent metal The material of the component that is to be welded.

Penetration The depth of the fusion zone.

Run (weld) This is the length of the weld pool for one pass of the filler rod or electrode.

Scale drawing This is a **drawing** that shows a real object with accurate sizes reduced or enlarged by a certain amount (the **scale**). The **scale** is shown as the length in the **drawing**.

Spatter Globules of molten metal thrown out during welding.

Weld pool The pool of liquid (molten) metal formed during fusion of parent and deposit metals.

Welding sequence (direction) The order and direction in which welds and runs are made.

Index

Great Clarendon Street, Oxford, OX2 6DP, United Kingdom

Oxford University Press is a department of the University of Oxford.

It furthers the University's objective of excellence in research, scholarship, and education by publishing worldwide. Oxford is a registered trade mark of Oxford University Press in the UK and in certain other countries

First published in 2019

British Library Cataloguing in Publication Data
Data available

978-0-19-839561-4

10 9 8 7 6 5 4 3 2

Paper used in the production of this book is a natural, recyclable product made from wood grown in sustainable forests.

The manufacturing process conforms to the environmental regulations of the country of origin.

Printed in Great Britain by CPI Group (UK) Ltd., Croydon CR0 4YY

Acknowledgements
The publisher would like to thank the following for permissions to use copyright material:

Cover image: Highwaystarz-Photography

Artwork by QBS

p94: Colorfulworld86/Shutterstock

Although we have made every effort to trace and contact all copyright holders before publication this has not been possible in all cases. If notified, the publisher will rectify any errors or omissions at the earliest opportunity.